On the Design of
Game-Playing Agents

Synthesis Lectures on Games and Computational Intelligence

Editor
Daniel Ashlock, *University of Guelph*

Synthesis Lectures on Games & Computational Intelligence is an innovative resource consisting of 75-150 page books on topics pertaining to digital games, including game playing and game solving algorithms; game design techniques; artificial and computational intelligence techniques for game design, play, and analysis; classical game theory in a digital environment, and automatic content generation for games. The scope includes the topics relevant to conferences like IEEE-CIG, AAAI-AIIDE, DIGRA, and FDG conferences as well as the games special sessions of the WCCI and GECCO conferences.

On the Design of Game-Playing Agents

Eun-Youn Kim and Daniel Ashlock

ISBN: 978-3-031-00991-4   paperback
ISBN: 978-3-031-02119-0   ebook

DOI 10.1007/978-3-031-02119-0

A Publication in the Springer series
SYNTHESIS LECTURES ON GAMES AND COMPUTATIONAL INTELLIGENCE

Lecture #2
Series Editor: Daniel Ashlock, *University of Guelph*
Series ISSN
Print 2573-6485   Electronic 2573-6493

On the Design of Game-Playing Agents

Eun-Youn Kim
Hanbat National University

Daniel Ashlock
University of Guelph

SYNTHESIS LECTURES ON GAMES AND COMPUTATIONAL INTELLIGENCE #2

ABSTRACT

Evolving agents to play games is a promising technology. It can provide entertaining opponents for games like Chess or Checkers, matched to a human opponent as an alternative to the perfect and unbeatable opponents embodied by current artifical intelligences. Evolved agents also permit us to explore the strategy space of mathematical games like Prisoner's Dilemma and Rock-Paper-Scissors. This book summarizes, explores, and extends recent work showing that there are many unsuspected factors that must be controlled in order to create a plausible or useful set of agents for modeling cooperation and conflict, deal making, or other social behaviors. The book also provides a proposal for an agent training protocol that is intended as a step toward being able to train humaniform agents—in other words, agents that plausibly model human behavior.

KEYWORDS

mathematical games, game-playing agents, automatic training of agents, representation in evolutionary computation

Contents

Preface

One of the pillars of experimentation is proper controls. This book summarizes a number of publications that demonstrate a lack of proper controls in experiments where game-playing agents are being trained. The central issue treated in the book is one of representation—the way that the agents are encoded for use in software. However, a number of other issues also arise as examples of insufficient control. These include the amount of computational or informational resources granted to the agents as well as some fairly basic algorithmic details of the training algorithms themselves.

Agent-based simulation has the potential to enable complex, nuanced simulation of human behavior, perhaps rising to the level of prediction. The authors have published dozens of articles in the direction of realizing this vision, almost all of them pointing out factors that cause trouble when they are not properly controlled.

In addition to enumerating a long list of pitfalls for the agent-based modeler, this book also serves as an introduction to the process of designing a representation for agents to be used in a simulation. Examples of the types of resources that can be made available to an agent abound from very simple agents that encode probability distributions on a set of moves to quite complex agents that have evolvable hormonal organs that drive their behavior.

The book also examines the issue of the simulated environment in which the agents are trained. An environment consisting of only other agents can be quite complex, but adding elements like recognition tags, records of past actions, and geography to the environment permit modeling of more complex and more plausibly human situations.

This book is not a finished work on how to design agent-based game-playing agents; we are simply not there yet. Rather, it is a progress report as to the research community with two goals: (1) to help others avoid traps that the authors have fallen into and (2) to encourage others to join us in the enterprise of refining the technique of agent-based modeling to make them more useful and reliable.

Eun-Youn Kim and Daniel Ashlock
July 2017

Acknowledgments

This book draws on joint publications with many collaborators. The authors would like to thank Nicole Leahy who collaborated on the original work on agent representation. Wendy Ashlock collaborated on the analysis of agents trained in the presence of noise, contributing many valuable ideas, as well as providing valuable insight throughout the project. Mark Joenks invented ISAc lists, one of the more versatile representations used in these investigations. Simon Lucas and Spyridon Samothrakis sparked the ideas of morphing games as well as collaborating in the investigation of the resulting game spaces. Gary Greenwood is the principle researcher in the work reported here on the Snowdrift game and has been an invaluable collaborator on the generalizations of Divide-the-Dollar reported here. The material on tags and the fine-grained treatment of non-local adaptation are adapted from a joint paper based on Brad Power's Masters thesis. Colin Lee, Justin Shonfeld, and other students and collaborators that have worked with the authors over the years have helped us shape and advance the research agenda that led to this book.

Eun-Youn Kim and Daniel Ashlock
July 2017

CHAPTER 1

Introduction

This book is about game theory, but from a unique perspective. In the economic applications of game theory, the goal is to solve the games, to find the Nash equilibria, and to find the evolutionary stable strategies. These properties of a game are supposed to permit the games, functioning as models of behavior, to make predictions about the behavior of agents that are engaged in the games. The problem is that human agents usually don't arrive at the predicted points. Even artificial agents with a modicum of complexity don't arrive at the stable points predicted by theory.

The famous statistician George E. P. Box observed that "All models are wrong; some models are useful." With this quote as a starting point, this book will make the case for an agent-based—as opposed to purely rational—set of techniques for experimenting with game theory. The hypothesis that this book attempts to support is that purely mathematical models of game theory are too simple to be useful in many circumstances, and that agent-based simulations can carry the load of prediction at least a little farther toward success.

This attempt to present and support alternative agent-based technology, while a goal, is *not* the main point of the book. In a series of over a dozen papers the author and her collaborators have found serious flaws in the methodology used to train game playing agents. The first and most central issue in this book is *representation*. The representation of an agent is the way it is encoded for use in the computer. Published work summarized here has examined may ways of representing agents to play a very simple game, the Iterated Prisoner's Dilemma. These include finite state machines, artificial neural nets, various forms of evolvable computer code, look-up tables, digital logic formulas—both trees and circuits—as well as stochastic encodings like simple Markov chains. These representations are described in detail in Chapter 2. The punchline is that the outcome of a very simple experiment that trains Prisoner's Dilemma agents could have its outcome not only affected but essentially *controlled* by the choice of how to represent the game-playing agents.

Another topic introduced in Chapter 2 is that of *fingerprinting*. This is a technique for broadly characterizing the behavior of an agent no matter how it is represented. A fingerprint computes the functional form of the agent's average payoff against an infinite collection of opponents, based on parametric descriptions of those opponents, giving a detailed signature of the way the agent plays. This representation independent signature is invaluable for documenting when some change in the agent training system impacts the character of the agents that emerge.

The recognition that representation of agents had a huge impact on agent behavior is a settled issue—the mechanisms by which representation acts to change behavior are subtle and still under investigation. Once it was noticed that representation had such a large influence, in effect representing a critical and uncontrolled variable in many publications, a very natural question is: *What else has not been adequately controlled?*

In Chapter 3, a number of secondary factors that also lack proper controls are examined. These include the payoff matrix (most games have a range of permissible payoffs), the amount of resources made available to an agent, and the algorithmic details of how the agents are trained. All of these factors are found to make significant differences in the same basic experiments as prisoner's dilemmas.

At this point in the narrative, a body of research has documented that work on the Iterated Prisoner's Dilemma suffers from a shortage of proper controls of a large number of factors—which rather begs the question of what happens in other games. Chapter 4 examines these issues in the context of other games. The concept of morphing between games is used—in this instance between Rock-Paper-Scissors and a novel game called Coordination Prisoner's Dilemma. A far more complex game, Divide-the-Dollar, is also examined. The chapter ends by examining one of the other classical games, the Snowdrift game, that can be used in a similar fashion to the Prisoner's Dilemma for game theory research.

Much of the economic research uses a very simple representation, that of a *mixed strategy* which places a probability distribution on the moves of the game and then seeks to optimize payoff as a function of those moves. This is a beautiful, mathematically idealized system that manages to ignore memory, mood, and other forms of conditioning that affect human behavior. In the context of this model, noise—errors in either intended play or the interpretation of another agent's play—are not difficult to deal with. Since the system already deals in expected values and probabilities, noise is a simple added term in the probability equations.

When more complex representations are used noise can have an enormous impact and, as such, deserves much closer study. In Chapter 5, we examine the impact of noise on the finite state agents. Results include greater agent robustness and more stable behavior over time. Based on the change of fingerprints, this chapter also introduces the new metric of *evolutionary velocity* which permit understanding of the impact of noise on the evolving systems.

Chapter 6 contains a prototype of a checklist for classifying representations. The set of classifying principles given in this chapter are intended as a first draft and community comment is welcomed. This chapter also introduces the notion of computational emotions. These are very complex representations in which the agents are first given and then permitted to evolve their own internal state engines that have been variously called computational emotions, digital hormones, and long-term memories.

1.1   PRISONER'S DILEMMA AND OTHER GAMES

The Iterated Prisoner's Dilemma (IPD) [38] is a classic of game theory. The game is often explained with the following story. Two miscreants have been captured by a law officer. He thinks they have pulled off a large burglary but only has evidence for the lesser crimes of trespass and property damage. The prisoners are placed in different rooms and each offered the same deal—the law will be lenient with them if they testify against their partner. If one of the prisoners decides to *cooperate* then he will maintain silence, *defection* consists of betraying his partner in crime. There are four possible outcomes.

- Both prisoners cooperate. In this case they serve the smaller penalties for the crime they are clearly guilty of and, subsequently, divide the loot. Both prisoners receive the *cooperation* payoff, C.

- One prisoner defects against the other (this can happen two ways). In this case, the cooperating prisoner gets the book thrown at him both for being guilty of the greater crime and also for being uncooperative. The defecting prisoner is granted immediate release and also may well manage to retain some of the proceeds of the burglary. The defecting prisoner receives the *temptation* payoff T and the other receives the *sucker* payoff S.

- Both prisoners defect. In this situation, both prisoners are penalized for the greater crime, but their penalty is mitigated to some degree by their cooperation with the law. Both prisoners receive the defection payoff D.

This game, like all the games studied here, is *simultaneous* in that the players play at the same time, or at least without knowledge of the other player's action while deciding how to play. In order to be considered Prisoner's Dilemma, the four payoffs must obey two inequalities:

$$S \leq D \leq C \leq T \tag{1.1}$$

$$S + T \leq 2C \ . \tag{1.2}$$

The first of these simply places the payoffs in a sensible order while the second requires that alternating cooperation and defection not be a better choice, on average, than cooperation. A common set of payoff values, one used in the experiments presented in this book when no other values are specified, is $S = 0$, $D = 1$, $C = 3$, and $T = 5$. One place where other payoffs are used is in Chapter 3 where the impact of changing the payoffs—within the values allowed by Equations 1.1 and 1.2—is shown to have an impact on which agents are located by the training algorithm.

The purpose of using numerical payoffs is to make Prisoner's Dilemma into a mathematical game. When the work expands to additional games, this consists of simply changing (and explaining) the payoff values. With this formalism in place we may specify new games by simply naming the moves and then specify the payoff matrix that lists the result for each pair of plays that can be made by the agents.

1.2   DIGITAL EVOLUTION: EVOLVING GAME PLAYERS

The key technology used to train agents is *evolutionary computation* [3]. These algorithms use a simplified digital form of the theory of evolution to perform optimization and can be adapted to agent training in a transparent fashion. These are the steps, remember that higher scores are presumed to be better.

1. Obtain a population of agents, e.g., generate them randomly.

2. Repeat.

3. Have the agents play the game in some fashion.

4. Pick agents with a direct scoring bias to reproduce.

5. Generate new agents from those picked to reproduce.

6. Place new agents in the population, possibly displacing old ones.

7. Until agent training is complete.

This training technique is given at a very high level and more details are filled in for the experiments in subsequent chapters. One of the results reported in Section 3.3 is that changing these algorithmic details have a substantial impact on the type of agents that emerge from training.

Each of the steps in the training algorithm has may different potential forms. Instead of generating the initial population randomly, for example, they could be designed with a heuristic or obtained as the results of another training algorithm.

In many of the experiments presented here the agent's quality is evaluated in a round-robin tournament. This is a choice, one that leans toward stability and away from discovery. Having small, randomly selected groups of agents play one another to estimate agent quality would permit the agent training algorithm to cast a much wider net for novel strategies, for example.

When we pick agents to reproduce, a bias toward quality is required to ensure that one of the primary requirements of evolution be met: that superior individuals reproduce. This comes with caveats—in nature success at reproduction is the currency of quality while in artificial evolution we hijack this and make our goals the definition of fitness. The issue of the strength of the pro-quality bias is important. A very strict bias permits evolution to settle on a relatively good design without performing a broader search that could have found better designs. This is called *premature convergence to a local optima* though a mathematician would not recognize this process as convergence in the classical sense at all.

When generating new agents from those picked for reproduction, it is typical to take pairs of parents, blend them to create children (this is called *crossover*), and then to tweak the children with a few random changes (this is called *mutation*). The search power of crossover and mutation, which are examples of *variation operators*, is highly dependent on exactly how the structures being evolved are stored in the computer. Given a choice of structure, many choices remain about how to implement the child generation process. *Brood recombination*, for example, generates a large swarm of children, choosing only the one or two best. These issues are beyond the scope of this text, but the choices made in the experiments presented are careful specified.

1.2.1   OPTIMIZATION VS. CO-EVOLUTION

The original application of evolutionary computation was to the optimization of solutions to fixed problems. This meant that, when we are determining the quality of structures used to drive selection, reproduction, and ultimately evolution, we are searching a fixed landscape that is in effect the graph of the function we are using to estimate quality. This goes out the window when we train agents to play games by having them play against one another.

The quality of an agent is measured by having it play other agents from those currently present in the population. Since this agent population is updated in each iteration of the algorithm, this means that the standard of quality changes along with the agent population. The standard form of evolution, against a fixed standard, is called *evolution*. The evolution of game-playing agents against a changing population of other game-playing agents is one type of *co-evolution*. As we will see in Chapter 3, changing the rate at which the agent population is updated is enough to change the population of agents that result from evolution.

The evolving agent population is a form of discrete dynamical system; it can follow complex trajectories through the space of possible populations of game-playing agents. The stochastic peregrinations of a population of agents through the space can easily be deflected by algorithmic details. Worse still, the very nature of the space is implicitly specified when the agent representation is chosen. These factors may illuminate why the impact of representation was more easily noticed in a games environment.

It is worth revisiting the comments in the last section about the functioning of digital evolution. There it is stated that there is a requirement, for evolution to take place, that superior individuals reproduce. In a co-evolving population of game-playing agents the notion of superiority is purely local to the current population. This opens the door to much stranger possibilities such as the periodic re-emergence of a strategy as some cycle in the strategy space brings it back to superior status.

Another factor is that, since new opponents are arising all the time, there is a substantial premium placed on being able to react quickly to changes in the environment. This means that keeping previously good strategies "nearby"—where the definition of distance is related to the probability of recovering the behavior via the action of the variation operators—is a behavior that is placed at a substantial premium. A complex secondary game is played in the genetics of the evolving agents.

1.3 FROM CHECKERS THROUGH THE RISE OF THE GO MACHINES

The history of digital game playing has been, for the most part, the history of teaching machines to play games that are of interest to humans. One of the earliest stories about a machine playing games was the Turk, a purported Chess-playing automaton that was actually run by a small person concealed inside it.

An early machine that played Tic-Tac-Toe was "Bertie the Brain." This automaton was constructed in 1950 for the Canadian national exposition. Since a young child can master flawless strategy for Tic-Tac-Toe (Bertie could lose), this example cannot really represent the first instance of intelligent machine game playing.

The first serious machine learning effort that attempted to master a game was the brainchild of Arthur Samuel who tackled *checkers*—also called *draughts*. Samuel introduced the formalism of searching a tree of game configurations as a way of formally presenting a game for digital computation and invented the first useful tree-pruning algorithm, *alpha–beta pruning*. This technique permits the computer to ignore parts of the search tree for the game when those parts of the tree are masked by really bad moves that would prevent them being considered. While this pruning technique has a huge impact, Samuel himself considered the pruning method so obvious he did not highlight it as a distinct achievement. Checkers research continued until a program called Chinook became the world champion at checkers and the game was declared "solved," meaning that the full game tree is now known.

The human world champion at Chess, Gary Kasparov, was defeated by a machine for the first time in 1997. The see-sawed back and forth for a time with algorithmic players becoming clearly superior around 2006. Human masters are now defeated by code that runs on mobile phones. This defeat of humans in what had been a standard test of superior intelligence led people to focus on a much harder game, *Go*.

It is quite difficult to determine how hard a game is for human players as we don't really know how they play games. When it is said "Go is harder than Chess" this evaluation is performed by looking at the game tree formalism of Samuel. The size of the tree of all possible moves for the game is estimated. That tight, quantitative measure makes Go about 10^{20} times harder than Chess. This incredible difference arises from the fact that chess removes pieces over time while Go just continues to fill in a complex board. As the game progresses, the number of possible next moves must shrink for Chess. The possibility space of Go starts much larger— compare the board sizes—and a game of Go usually concludes before the board fills in enough to make the set of next moves small again. It is worth remembering that size is probably not a good direct measurement of difficulty.

Recently, it was predicted that Go would require another decade to fall to the machines— but it only took a little more than a year after that prediction that ten years would be required. A technology called *deep learning*, which had been discussed for some time but which recently become computationally practical, was united with a remarkable search technique called *Monte-*

Carlo tree search. Monte-Carlo tree search was, in fact, invented to deal with Go. The big difference between Monte-Carlo tree search and other tree search techniques is that it performs biased sampling of the game tree, rather than trying to complete a certain number of layers of the game tree.

These techniques are not the focus of this book, but they are included as pointers into another rich part of games as a research area. They do, however, raise an interesting issue. The focus of this book is putting games that model political and economic behavior on a firmer footing. If problems with agent-based models that predict human behavior can be reduced or removed, are there implications in the recent, massive success of game playing technologies for placing our behavior modeling capabilities at a new level?

1.4   WHY ARE THE SIMPLE GAMES SO HARD?

A standard game-theoretic analysis of a simultaneous two-player game may yield a list of equilibria which form the game-theoretic predictions about where humans in the situation modeled by the game may end up. As noted previously, the author does not yet consider this a reliable technology. In comparison with the victory of computational tools over human skill in even the game of Go, why are simple games like Prisoner's Dilemma so hard to get right?

The obvious answer is that the games of this sort are such simple models, and require such strong assumptions, that they fall below the level of predictive power required to model human behavior. There is a more subtle issue of difficulty that the author seeks to address. Even if we accept the simplicity or games like Prisoner's Dilemma and do no more than generalize the type of play strategies permitted, the research summarized here shows that we end up in a quagmire of choices and assumptions.

A standard representation for a game playing agent in traditional game-theory research is a *mixed strategy*, by which is meant a probability distribution on the moves of the game without even the memory present in a Markov chain. These representations also make the assumption that the number of actors is infinite; this is the *infinite population* assumption. The use of this representation and these assumptions is well motivated: it permits the researcher to get formally supported results about the game via calculations and mathematical proof. One problem with the approach is that human beings, the targets of the model, are probably not memory-free random number generators, nor do they form an infinitude of individuals.

The author and her collaborators have examined social structures, simulated emotions, geographical constraints, and a number of other generalizations to the basic game theoretical model that attempt to capture more of the richness of human experience. In the end, all of this must wait, however, on understanding and stabilizing the problems that arise from simply using more complex digital representations of agents that play the simplest versions of simultaneous two-player games.

The difficulty in these "simple" games arises from the fact that they are played by interacting populations of agents which form a complex structure in their own right. Another factor is that when play is open-ended, then so is the corresponding space of strategies. Finally, in order for the agent-based versions of these games to be useful, they must deal efficiently with irrationality, bounded rationality, and random chance. This book solves only some of the many problems it identifies. Its major contribution is to place some warning signs about road-side obstacles on the journey toward an ability to model human behavior.

CHAPTER 2

The First Place Where Trouble Arose

This chapter chronicles the first observations that experiments which evolved agents to play simple games were performed while neglecting a critical class of controls. A central issue in evolutionary computation is the representation issue. To reiterate—this is the issue of *how* a structure is realized in digital form to permit optimization—evolutionary or otherwise—to operate on it.

Begin with a very simple example. Suppose that you are optimizing a real function with 20 variables. Would it be more sensible to evolve a gene that is an array of 20 real numbers or a gene that is a 960-bit string that codes for real numbers using blocks of bits to code for the individual real parameters? Should the crossover operator respect the boundaries of real numbers or be allowed to divide the structure in the middle of a real number? Would a representation that specifies how to locate the set of 20 parameters be more effective? These questions are being asked about an optimization domain in which the measure of quality is changing.

When we turn to the domain of game-playing agents that are evaluated against one another, the changing nature of the quality measure or fitness function makes these issues harder to understand. This chapter summarizes the initial experiments performed on changing the way agents are represented, gives a representation-independent tool (fingerprinting) for assessing impact of changing representation, and then documents that changing representation in fact dominates the outcome of a standard emergence-of-cooperation experiment.

2.1 REPRESENTATION—WHAT IS IT?

The IPD is a widely used computational model of cooperation and conflict. Many studies report emergent cooperation in populations of agents trained to play Prisoner's Dilemma with an evolutionary algorithm. It is of particular importance to note that evolutionary algorithms do not just optimize for high scores. The use of evolution also imposes secondary implicit fitness criteria related to the ability of a strategy to survive the reproduction process. Since different representations both encode somewhat different sets of strategies and impose different patterns of inheritance on evolving populations, there is the potential for behavior of the system to depend on representation. Understanding this "pattern of inheritance" is critical for certification of any application of evolutionary algorithms to behavioral simulation.

There are many possible motives for performing an evolutionary IPD study. These motives inform the degree to which the researchers need to control for their choice of representation. If the goal is to simply discover new strategies for the IPD, then no controls are needed; the researchers will distill their output down to clearly stated strategies in any case. At the other extreme, evolutionary studies can be used to suggest game-theoretic theorems that the researcher would like to prove. The sensitivity of evolutionary algorithms to the choice of representation means that computational simulations are not likely to be representative of general truths about the games under study. Rather, they are samples of a complex interaction between game, representation, and the dynamics of evolution.

Definition 2.1 A **representation** is the data structure used to store an agent together with the variation operators applied to the structure during reproduction.

The definition used in this book is motivated by the following structural observations. Suppose that we have settled on a data structure for storing agents. The points in the space of agents are the different ways the details of this data structure may be filled in. The *connectivity* or *topology* of the space is generated by the action of the variation operators. The evolutionary search for higher scoring strategies proceeds by mutation and crossover, and so these variation operators connect the current population to the next one and, by implication, the entire space.

It is worth noting that several of the representations used in this book, e.g., Boolean formulas, look-up tables, and artificial neural nets, encode precisely the same set of strategies. In spite of this, they exhibit substantially different levels of cooperative behavior in the experiments performed. This suggests that representation cannot be considered separately from the variation operators—the presence of a strategy in an encoding is not sufficient to ensure it will be discovered.

To aid in discussion of the results or the experiments evolving Prisoner's Dilemma playing agents, Table 2.1 gives a number of example strategies for playing the IPD.

2.1.1 FINITE STATE MACHINES, DIRECT REPRESENTATION

A finite state machine is a finite collection of states and possible inputs, where each state and input value has a transition to another state and an output associated with it. The states are internal markers used as memory, like the tumblers of a combination lock that "remember" whether the user is currently dialing in the second or third number in the combination.

Example 2.2 In IPD, the game strategy Tit-For-Two-Tats which is an implementation of the defects only if its opponent has defected on the last two actions is specified by a finite state machine, as in Figure 2.1. The finite state machine from Figure 2.1 can be into an array of integers in the structural manner. First, we strip a finite state machine down to the integers that describe it (setting $C = 0$, $D = 1$) as in Figure 2.1. To find the structural grouping gene we simply read the stripped table from left to right, assembling the integers into the array: 0001001100.

Table 2.1: Examples of Prisoner's Dilemma strategies

Always Cooperate (AllC) This strategy always plays C.

Always Defect (AllD) This strategy always plays D.

Fortress-3 (Fort3) This strategy is an example of a strategy that uses a password. If the opponent defects twice in a row (the password) and cooperates thereafter, then Fortress-3 will cooperate. Any deviation from this sequence resets the need to defect twice. Fortress-3 was first defined in [37] and is an example of a strategy that only arises after substantial evolution has taken place.

Majority (Maj) This strategy returns a play equal to the majority of its opponent's plays, breaking ties in favor of cooperation. Majority has no finite state representation.

Pavlov (Pav) The strategy, Pavlov, plays C as its initial action and cooperates thereafter if its action and its opponent's actions matched last time. A minimal finite state implementation of Pavlov is shown in Figure 2.2.

Periodic CD (PerCD) This strategy cooperates and defects on alternate moves no matter what its opponent does.

Psycho (Psy) The strategy, Psycho, chooses D as its initial action and then plays the opposite of its opponent's last action.

Punish Once (Pun1) The strategy punishes the first defection by an opponent but cooperates thereafter.

Random (Rand) The Random strategy simply flips a fair coin to decide how to play. Random has no finite state representation.

Ripoff (Rip) This strategy alternates cooperation and defection until its opponent defects for the first time. On the round after this defection, it cooperates and then plays tit-for-tat thereafter.

Thumper (Thmpr) This strategy cooperates initially. If its opponent defects, then it defects on the next two moves; if its opponent's second move after defection is cooperate, it continues cooperating; otherwise it defects twice as before.

Tit-for-tat (TFT) The strategy, tit-for-tat, plays C as its initial action and then repeats the other player's last action.

Tit-for-two-tats (TF2T) This strategy defects only if its opponent has defected on the last two moves.

Tit-for-three-tats (TF3T) This strategy defects only if its opponent has defected on the last three moves.

Two-tits-for-tat (TTFT) This strategy defects on the two moves after its opponent defects, otherwise it cooperates.

The finite state machines used here are 16-state Mealy machines; responses to an opponent's actions are encoded in their transitions. State transitions are driven by the opponent's last action. Access to state information permits the machine to condition its play on several of its opponent's previous moves. The machines are stored as linear chromosomes; the first state in the chromosome is the initial state. The crossover operator used is two-point crossover of the list of states. Crossover preserves whole states. The point mutation operator changes a single state transition, the initial state, the initial action, or an action associated with a transition. The object to change is selected uniformly at random, and then a valid value for the state transition or action is selected, also uniformly at random.

Tit-for-Two-Tats				Tit-for-Two-Tats	
Initial Response: C				0	
Initial State: 1				0	
State	If D	If C			
0	$C \to 1$	$C \to 0$		01	00
1	$D \to 1$	$C \to 0$		11	00

Figure 2.1: A finite state implementation of the strategy *Tit-For-Two-Tats*. This machine is shown both in human readable form and as it might be encoded in bits.

Always Cooperate		
Initial Response: C		
Initial State: 1		
State	If D	If C
1	$C \to 1$	$C \to 1$

Always Defect		
Initial Response: D		
Initial State: 1		
State	If D	If C
1	$D \to 1$	$D \to 1$

Tit-for-Tat		
Initial Response: C		
Initial State: 1		
State	If D	If C
1	$D \to 1$	$C \to 1$

Tit-for-Two-Tats		
Initial Response: C		
Initial State: 1		
State	If D	If C
1	$C \to 2$	$C \to 1$
2	$D \to 2$	$C \to 1$

Pavlov		
Initial Response: C		
Initial State: 1		
State	If D	If C
1	$D \to 2$	$C \to 1$
2	$C \to 1$	$D \to 2$

Figure 2.2: Finite state machine tables for some examples of common Prisoner's Dilemma strategies.

2.1.2 FINITE STATE MACHINES WITH A CELLULAR REPRESENTATION

The cellular representation [64] for finite state machines modifies an initial one-state machine with a series of editing commands. The data structure used for this representation is a string of editing commands. The initial machine has an initial action of cooperation. The initial machine returns its input as its next output, echoing the other player's actions. This machine is called an echo machine. For the IPD the echo machine is an encoding of tit-for-tat, a strategy that cooperates initially and repeats its opponents last action thereafter. It is possible that the use of tit-for-tat as the initial state of the editing system might bias the system in favor of this strategy. The data reported in Table 2.5 show that tit-for-tat was, in fact, relatively rare in the populations evolved with this encoding and so there is apparently little need for concern. The rules for modifying the initial echo machine into the machine encoded by the cellular representation are given in Table 2.2. In order to execute these rules, it is necessary to use structures not found in the final finite state machine. The first of these structures is the current state pointer. This pointer designates the state to which an editing command is applied. This pointer may be moved, and it may drag a transition arrow with it. This is done by using the pin command to associate one of the transition arrows out of the current state with the current state pointer. Later, the release command releases the head of the "pinned" transition arrow on the current state as designated by the position of the current state pointer. If a pin command is executed while there is already a pinned transition arrow, then it is implicitly released at the current state. While pinned, the transition arrow's head moves with the current state pointer. A pinned transition arrow is also implicitly released if still pinned at the end of the construction of a machine. The second structure beyond the nominal finite state machine is a collection of pointers connecting each state in the machine to the state that was duplicated to create it. These pointers are used to execute the Ancestor command. The first state has no ancestor pointer, and the ancestor command is ignored if it is executed while the current state pointer is pointing to the first state. Notice that only 2 of the 13 rules, D_C and D_D generate a new state. This means the number of states in a machine is equal to the number of D_n rules in the cellular encoding plus one.

Table 2.2: Editing rules for the cellular representation. The symbol n can be a cooperate or a defect.

Editing Command	Effect
B (Begin)	Flip the initial action.
F_N (Flip)	Flip the response associated with the transition for input n out of the current state.
small M_n (Move)	Move the current state to the destination of the transition for input n out of the current state.
D_n (Duplicate)	Create a new state that duplicates the current state as the new destination of the transition for input n out of the current state.
P_n (Pin)	Pin the transition arrow from the current state for input n to the current state. It will move with the current state until another pin command is executed.
R (Release)	Release the pinned transition arrow, if there is one.
I_n (Square)	Move the transition for input n out of the current state to point to the state you would reach if you made two transitions associated with n from the current state.
A (Ancestor)	Move the current state to the state that was duplicated to create the current state or do not move it if the current state is the initial state.

2.1.3 BOOLEAN FOMULAS

The Boolean formulas in this study use the operations given in Table 2.3. Cooperation was encoded by the Boolean value false while defection was encoded by true. The delay operation holds a value for one evaluation cycle before passing it on, returning a cooperate when first accessed. The delay was disabled in one of two sets of evolutionary runs. The data structure and variation operators used are described in [40]. The type of evolutionary computation used to evolve the Boolean formulas is termed genetic programming. Initial populations of trees were generated recursively by selecting a root node at random from the available operations and then dividing the remaining nodes between the arguments of that operation. These remaining operations were used to generate trees in the same manner, terminating with single node trees that are terminals. The number of nodes assigned to each argument is selected uniformly at random from the choices that leave at least one node for each argument. This tree-generation process considers the number of nodes remaining when deciding which operations are available; a two node tree is a unary operation whose argument is a terminal, for example. The formulas in the initial population were generated with 6 operations and terminals but were permitted up to 12 during evolution. Note that, given the available operations, 12 nodes is more than the number required to permit all three-input Boolean functions.

Table 2.3: Terminals and operations used in the tree representation of Boolean formulas

Name	Arity	Definition
T	0	Constant-logical true.
F	0	Constant-logical false.
X_i	0	Input, opponents previous actions, i = 1, 2, 3.
Not	1	Logical not, inverts truth value.
Del_*	1	Delay line: returns true when first called and its last input thereafter.
Say	1	Repeats its input.
And	2	Logical and, true only if all inputs true.
Or	2	Logical or, true if any inputs are true.
Nand	2	Logical nand, inverted and.
Nor	2	Logical nor, inverted or.
Xor	2	Logical exclusive or, logical inequality.
*Used in one set of simulations, absent in the other.		

We use subtree crossover and subtree mutation in addition to a novel variation operator called chop in Section 2.3. Subtree crossover selects a node uniformly at random in each tree. The subtrees rooted at those nodes are exchanged. The subtree mutation operator selects a node uniformly at random. The subtree rooted at that node is then replaced with a new subtree with the same number of nodes generated in the same fashion as the trees in the initial population. When variation operators create trees with more than 12 nodes, the chop operator is applied. A randomly selected argument of the root node of the tree is promoted to be the root node, iteratively, until the tree had no more than 12 nodes.

2.1.4   FUNCTION STACKS

A function stack is a representation derived from Cartesian Genetic Programming [86]. The parse tree structure used in genetic programming is replaced with a directed acyclic graph with a form of time-delayed recurrent link. An example of a one-node function stack is shown in Figure 2.3. The vertices of this graph are stored in a linear chromosome. Each node specifies a binary Boolean operation, an initial output value for the recurrent output of the operation, and two arguments. The available Boolean operations are: And, Or, Nand, Nor, Xor, and Equality (Nxor). The available arguments are the Boolean constants true and false, the opponent's last action, the output of any Boolean operation with a larger array index than the current one, or the output for the previous time step of any Boolean operation in the function stack. This latter type of argument is called a recurrent link. The function stacks have a "feed forward" topology, a directed acyclic graph save that the output of every operation in the previous time step is available to every operation as a potential argument. This yields a form of short-term memory or time-lagged recurrence. The initial output values for the recurrent links of each node are required to give the value used for that link on the first time step.

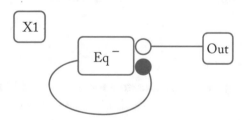

Figure 2.3: A function stack with one node that implements the strategy Pavlov. Pavlov cooperates initially and cooperates thereafter if it made the same play as its opponent in the previous time step. The input X_1 is the opponent's last move. The black circle is the output associated time-delayed recurrent link while the output of the equality operation is the white circle. The minus superscript denotes the initial output value of the recurrent link; in this case it is false.

Allowing the function stack access only to the opponent's last move was done for the same reason as with finite state machines. The recurrent links give the function stack internal state information. This permits them to store the opponent's previous actions. This means that function stacks, while they nominally only see the opponent's last action, can evolve to use more past actions than strategies that condition their action on the opponent's last three moves like look-up tables.

The function stacks used in Section 2.3 have 16 nodes analogous to the number of states used in the finite state machines. The actual tradeoff of states for nodes is complex and unclear. The recurrent links yield 2^n available internal states in a function stack with n nodes, but these states are more difficult to access than the states of a finite state machine as they are distributed over the entire structure.

An agent's action is the output of the first node, also called the output node. During initialization, operations are selected uniformly at random from those available. Arguments are selected according to the following scheme: one argument in ten is a constant (true or false), one quarter of all arguments are recurrent links to another node, selected uniformly at random, the remainder of the arguments are either the output of nodes farther down the function stack or input variables. The probability that an argument will be an input variable (if it is not a memory or constant link) is directly proportional to the distance from the output node. Thus, the first node is most likely to reference the output of other nodes, and the arguments of the last node will be input variables if they are not memory links or constants. This linear ramp-up encourages a larger fraction of links from the output node to other nodes, either directly or indirectly.

The binary variation operator used on function stacks in Section 2.3 is two-point crossover of the linear chromosome of nodes. The point mutation operator chooses a random operation three eighths of the time, a random argument half the time, and an initial value for a node's memory one-eighth of the time. If an operation is selected, then it is replaced with another operation selected uniformly at random. If an argument is selected, then it is replaced with a valid argument selected according to the scheme used in initialization. If an initial memory value is selected, then it is inverted.

2.1.5   ISAC LISTS

The chromosome structure and variation operators for If-Skip-Action (ISAc) lists are described [33]. An ISAc list is a list of ISAc nodes. Each of these nodes contains four fields: a pair of indices into a data vector, an action, and a jump location. When a node is executed, the two data items indexed by the node are compared. If the first is larger than the second, then the node executes its action field. The action may be a null action, a Prisoner's Dilemma action (cooperate, defect), or a jump action. To execute a jump action, the ISAc list transfers execution to the node specified by the jump location; otherwise the next node executed is the next node in the list. The list is circular, executing the first node after the last one. Null actions serve as place holders permitting mutation to insert or delete actions.

The data vector of the ISAc list contains the opponent's last three moves, encoded 0 for cooperate and 1 for defect, as well as the constants 0, 1, and 2 for comparison. These three actions are initialized to zero before play. In a given session of play, an ISAc list starts executing nodes with its first node and continues executing nodes until it reaches a Prisoner's Dilemma action. This action is returned as the agent's move. The jump action permits ISAc lists to have loops, increasing the expressive power of the representation. It also creates the potential problem of infinite loops that do not generate Prisoner's Dilemma actions. To deal with infinite loops of this type an ISAc list is permitted to execute no more than 1,000 ISAc nodes during a given 150 round session of play. If it exceeds this allotment, then the moves are chosen uniformly at random for the rest of the play. This is a penalty, as random play is sub-optimal.

The ISAc lists used in Section 2.3 contain 256 nodes. This large number of nodes is predicated on the observation that many ISAc nodes are needed to simulate a state in a finite state machine. The binary variation operator is two-point crossover, operating on the list of nodes. The point mutation operator changes one of the four numerical parameters in a node to a valid value selected uniformly at random. Both the node and field to modify within the node are selected uniformly at random.

2.1.6 MARKOV CHAINS AND LOOK-UP TABLES

The Markov chains and look-up tables used in this study have a chromosome consisting of a table of eight probabilities of cooperation indexed by the eight possible ways the opponent's last three moves could be made. The entries in the table describing the Markov chains are real numbers in the range $[0, 1]$ storing the probability of cooperation. Look-up tables are like Markov chains save that the probability of cooperation is 1 (C) or 0 (D). Variation operators for these representations operate on the linear list of probabilities. The binary variation operator used is two-point crossover. The point mutation operator for Markov chains adds a uniform random number in the range $-0.1 \le x \le 0.1$ to one of the eight probabilities, reflecting at 0 and 1 to maintain valid values. Reflection moves the probability toward the interior of the valid range by the amount the mutation has taken it beyond a bound. Mathematically, reflection is:

$$R(x) = \begin{cases} -x & x < 0 \\ x & 0 \le x \le 1 \\ 2 - x & x > 1. \end{cases}$$

The point mutation operator for look-up tables inverts the action, exchanging cooperation and defection.

2.1.7 ARTIFICIAL NEURAL NETS

The type of artificial neural nets (ANNs) used in this study are described in [66]. They have a hidden layer of three neurons and a single output neuron. All neurons are 0-1 threshold neurons. The topology of the nets is shown in Figure 2.4. The chromosome for this representation consists of the 12 connection weights: 9 for the connections of the 3 input neurons to the hidden layer and 3 for the connections of the hidden layer to the output neuron. Two different neural representations were examined. In the first, termed the neutral ANN representation, the neurons had an activation threshold of 0. For the second, termed the cooperative ANN representation, the neurons had a threshold of 0.5, creating a 3:1 bias in favor of cooperation at the level of individual neurons. Neural connection weights were initialized in the range $-1 \le c \le 1$.

Opponents Moves	Probability of Cooperation		Opponents Moves	Palyler's Move
CCC	0.20		CCC	C
CCD	0.10		CCD	D
CDC	0.64		CDC	C
CDD	0.75		CDD	D
DCC	0.10		DCC	C
DCD	0.45		DCD	D
DDC	0.60		DDC	C
DDD	0.15		DDD	D
(1)			(2)	

Figure 2.4: An example of a Markov Chain (1) and a look-up table (2). (2) encodes the strategy tit-for-tat. Previous moves are given with the most recent on the right. The representations store only the right column of each table as a linear chromosome.

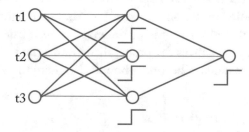

Figure 2.5: A feed-forward neural net with four 0-1 threshold neurons. The opponent's last three actions are used as inputs. A hidden layer with three neurons is used to drive a single output neuron.

To create a chromosome, the weights are placed into a linear array. The first nine entries are the input-to-hidden layer connections in groups of three associated with a given hidden neuron. The three remaining values are at the end of the chromosome and represent the connection of the hidden layer to the output neuron. The binary variation operator for this representation is two-point crossover. The point mutation operator adds a number selected uniformly at random in the range $-0.1 \leq x \leq 0.1$ to a connection weight also selected uniformly at random.

With these descriptions of a large number of different representations for game-playing agents made, if in a somewhat terse form, we are ready to harvest the main result of the chapter.

2.2 FINGERPRINTING—A TOOL FOR COMPARING ACROSS REPRESENTATIONS

Evolutionary game theory can use an evolutionary algorithm to generate a vast number of different game-playing agents in a short time. It is typically impractical to figure out *which* game playing agents were evolved for a number of reasons. Many representations for evolvable game-playing agents, e.g., neural nets or finite state automata, are capable of producing thousands of different encodings of the same strategy. In addition, evolved game playing agents are often cryptic and complex. The effort required for direct analysis of evolved structures varies from the time-consuming to the impractical. It is possible to create one-to-one representations that encode a small number of possible strategies, both solving the problem of understanding the evolved agents and depriving evolution of much of its scope to produce interesting strategies.

In this section we present *fingerprinting*, a representation-and-encoding independent method of identifying game playing agents. Fingerprinting is a technique for generating a representation-independent functional signature for a game playing agent. Fingerprints can be used to compare agents across representations in an automatic fashion. The theory of fingerprints is developed for software agents that play the IPD. The theory of fingerprints is developed to the point of supporting a rapid sampling-based algorithm for approximating fingerprints. This permits the rapid identification, sorting, and classification of game playing agents. While this study deals only with Prisoner's Dilemma, the technique extends to any simultaneous two-player game with a finite number of moves.

Now we gather together and extend the underlying theory of *fingerprints* for game-playing agents and applies fingerprints to a study of the types of Prisoner's Dilemma strategies that arise under evolution for different representations. Finite state machines, feed-forward neural nets, and look-up tables are found to sample the space of strategies in very different ways.

2.2.1 DEFINITION OF FINGERPRINTS

Fingerprinting was used in [37], with a finite state representation, to demonstrate that the strategies that arise have different distributions for different population sizes and in different epochs. The latter result, that strategies rare or absent at the beginning of evolution become common after thousands of generations of evolution, was surprising. In [21], fingerprints were used to demonstrate that the rate of appearance of several well-known strategies varied between a direct finite state representation for Prisoner's Dilemma playing agents, a cellular representation for finite state agents, and a new type of representation called a *function stack*, a modified form of Cartesian Genetic Programming [84].

The play of two finite state machines in the presence of noise can be represented as a *Markov process*. This allows the determination of an expected (average) score for any pair of strategies by standard techniques in stochastic processes [93]. We will use game-playing agents with strategies that incorporate parameterized noise to fingerprint other agents. The strategy used to evaluate other agents is called the *probe* strategy. The fingerprints have independent variables that establish the character of the noise and return a dependent variable that is the expected score of the agent being fingerprinted against the probe strategy. Noise represents probabilities of cooperating or defecting in spite of the move the probe strategy would normally have made. The fingerprint will thus be a map from probabilities, (x, y), of "irrational" (non-probe strategy) cooperation and defection, respectively, to a value, E, the expected score against the noisy agent.

Definition 2.3   If A is a strategy for playing the IPD, then $JA(A, x, y)$ (*Joss-Ann of A*) is a strategy which has a probability x of choosing the move C, a probability y of choosing the move D, and otherwise uses the response appropriate to the strategy A.

If S is the space of strategies for playing the IPD, then the Joss-Anne modification of strategies can be viewed as a function $A : S \times F \mapsto S$ where $F = \{(x, y)|x, y \in R, 0 \leq x + y \leq 1\}$ that yields a continuum of strategies.

Definition 2.4   A **fingerprint** $F_A(S, x, y)$ with $0 \leq x, y \leq 1$, $x + y \leq 1$ for strategy S with probe A, is the function that returns the expected score of strategy S against $JA(A, x, y)$ for each possible (x, y). The double fingerprint $F_{AB}(S, x, y)$ with $0 \leq x, y \leq 1$ returns the expected score of strategy S against $JA(A, x, y)$ if $x + y \leq 1$ and $JA(B, 1 - y, 1 - x)$ if $x + y \geq 1$. In this case A is the **lower probe** and B is the **upper probe**.

While the fingerprint function itself is often possible to find, it is the graph or the approximation of the graph that is often used in analysis. This is in part because a useful approximation of the graph of the function can be computed in cases where the analysis to find the actual function would be intractable. The concept of the double fingerprint was introduced to extend the fingerprint to the unit square in a natural fashion. A unit square is preferable because it is more easily manipulated by a computer, is more easily viewed by humans, and it uses paper more efficiently.

2.2.2   EXAMPLE FINGERPRINT COMPUTATION

The fingerprint of the strategy Pavlov, using tit-for-tat as a probe, is computed as an example. A minimal finite state representation of Pavlov is shown in Figure 2.2. To find a fingerprint, the first step is to construct a *Markov chain* for the two strategies involved. In the case of $F_{tit-for-tat}(Pavlov, x, y)$, the set of ordered pairs

$$\{(C1, D1),(C1, C1), (D2, D1), (D2, C1)\}$$

forms the (accessible) state space. A letter denotes an action. The numbers in a pair denote the internal (finite state machine) states of strategies Pavlov and tit-for-tat, respectively. Readers should verify for themselves that the given combinations of internal states and actions cover all attainable possibilities. Then, constructing the transition matrix, P, for the Markov chain is just a matter of putting the transition probabilities between the states in matrix form.

	$(C1, D1)$	$(C1, C1)$	$(D2, D1)$	$(D2, C1)$
$(C1, D1)$	0	0	y	$1 - y$
$(C1, C1)$	y	$1 - y$	0	0
$(D2, D1)$	$1 - x$	x	0	0
$(D2, C1)$	0	0	$1 - x$	x

Because this Markov chain consists of one finite communicating class, it has the stationary distribution π (see [93]), and it can be found by solving the equations $(P' - I)\pi = 0$ and $\sum \pi(i) = 1$. We obtain:

$$\pi = \left(\frac{y(1 - x)}{2y(1 - x) + x(1 - x) + y(1 - y)}, \frac{x(1 - x)}{2y(1 - x) + x(1 - x) + y(1 - y)}, \right.$$
$$\left. \frac{y(1 - x)}{2y(1 - x) + x(1 - x) + y(1 - y)}, \frac{y(1 - y)}{2y(1 - x) + x(1 - x) + y(1 - y)} \right).$$

With π in hand, computing the expected score can be completed by taking a dot product of π with the appropriate score vector $(S, C, D, T)'$ which gives corresponding scores for $((C1, D1), (C1, C1), (D2, D1), (D2, C1))$. This yields the fingerprint function

$$F_{tit-for-tat}(Pavlov, x, y) =$$

$$\frac{Sy(x - 1) + Cx(x - 1) + Dy(x - 1)^2 + Ty(y - 1)}{2y(x - 1) + x(x - 1) + y(y - 1)}.$$

In the case of the IPD, we usually score $S = 0, C = 3, D = 1$, and $T = 5$ so

$$F_{tit-for-tat}(Pavlov, x, y) =$$

$$\frac{3x(x - 1) + y(x - 1)^2 + 5y(y - 1)}{2y(x - 1) + x(x - 1) + y(y - 1)}.$$

2.2.3 FINGERPRINT RESULTS

Definition 2.5   Strategy $\mathcal{A}'$ is said to be the **dual** of strategy $\mathcal{A}$ if $\mathcal{A}$ and $\mathcal{A}'$ can be written as finite state machines that are identical except that their responses are reversed.

The strategies tit-for-tat and Psycho are examples of dual strategies. Tit-for-tat repeats its opponent's last choice. Psycho plays the opposite of its opponent's last choice. A strategy can be its own dual. For example, the strategy Pavlov is a self-dual strategy, as shown in Figure 2.2. Given the same input string, it generates reversed responses if its initial action is reversed.

Theorem 2.6   If $\mathcal{A}$ and $\mathcal{A}'$ are dual strategies, then $F_{\mathcal{A}\mathcal{A}'}(\mathcal{S}, x, y)$ is identical to the function $F_{\mathcal{A}}(\mathcal{S}, x, y)$ extended over the unit square.

This theorem shows that the dual strategy can be used to create a natural extension of the fingerprint into a dual fingerprint over the unit square and the following corollary gives us the metric with which game strategies can be classified.

Corollary 2.7   *If A and A' are dual strategies, then $F_{AA'}(\mathcal{S}, x, y)$ is infinitely differentiable over the interior of the unit square.*

A shaded plot of this function appears in Figure 2.6. The shading is a tool that permits the rapid identification of the fingerprint function in visualizations as demonstrated in [17].

Figure 2.6: A shaded plot of $F_{tit-for-tat}(Pavlov, x, y)$ for Prisoner's Dilemma. Lighter colors represent higher scores with black=0 and white=5. Shading is modified to show three important parts of the score space. The high shaded (blue) band represents scores within a narrow range of the score for mutual cooperation. The middle (red) shaded band similarly represents the score obtained by mutual random play. The low (green) shaded band marks scores near the score for mutual defection. The differential shading of this representation provides for rapid visual identification of fingerprints.

Theorem 2.8   *Look-up tables can be specified as finite state machines.*

Proof. The action of a strategy stored in a look-up table is determined by the opponent's previous finitely many moves. Let n be the number of the opponent's previous moves used to index the look-up table. We consider the opponent's previous moves as a string with length n, i.e.,

$x_1 x_2 \ldots x_{n-1} x_n$. Define the addition operator on strings to be concatenation. Construct a finite state machine in the following way. Assume the initial move is known. Construct $2^n - 1$ states. $2^{n-1} - 1$ states are transient states arranged as a binary tree needed for the first $n - 1$ moves. They lead to 2^{n-1} states in a single communicating class. To create these, use the first $n - 1$ digits of strings of a possible opponent's previous moves and denote them as $s[x_1 x_2 \ldots x_{n-1}]$. If the input is 0, then take transition from $s[x_1 x_2 \ldots x_{n-1}]$ to $s[x_2 x_3 \ldots x_{n-1} + 0]$ with the move corresponding to the string $x_1 x_2 \ldots x_{n-1} + 0$ in the look-up table. If the input is 1, then take transition from $s[x_1 x_2 \ldots x_{n-1}]$ to $s[x_2 x_3 \ldots x_{n-1} + 1]$ with the move corresponding to the string $x_1 x_2 \ldots x_{n-1} + 1$ in the look-up table. This creates the same directed graph as a De Bruijn sequence with $n - 1$-bit binary words [102] for the machine's sole communicating class. Clearly, this finite state machine mimics the behavior of the look-up table.   □

An example of a finite state machine constructed from a look-up table is given in Figure 2.7.

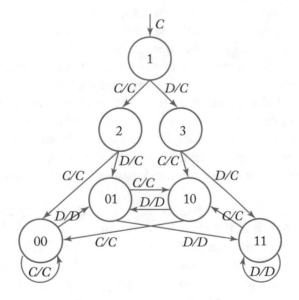

Figure 2.7: The finite state machine which specifies the look-up table in Figure 2.4.

Corollary 2.9   *Strategies encoded by Boolean formulas and feed-forward artificial neural nets can be specified by finite state machines.*

Proof. For both Boolean formulas and feed-forward neural nets, the structures specify a fixed mapping of the inputs to the outputs. This means that both representations encode a strategy that could be stored in a look-up table. Applying Theorem 2.8 demonstrates the corollary.   □

Notice that Markov chains encode strategies that cannot be specified by finite state machines. A strategy stored in a Markov chain has a continuous fingerprint that is not difficult to compute, but it can be one that cannot be realized by any finite state machine. This follows from the inability of finite state machines to simulate arbitrary probabilities of taking one of two actions. ISAc lists are a form of general-purpose computer program and so can easily encode non-finite-state strategies. It has been shown that function stacks and finite state machines can be used to simulate one another.

The facts about fingerprinting are used, for the most part, in later chapters, but are included here in part because their development included the realization that the strategy spaces of several of the representations studies were the same. This readies us for the most far-reaching result presented in this book.

2.3   THE UNCONTROLLED VARIABLE DISCOVERY

Many different sorts of evolutionary computation systems have been used to evolve agents to play the IPD. In [61], particle swarm optimization is used to co-evolve agents. Threaded finite state machines that permit multiple action threads based on the agent's internal state are used in [8]. Permitting Prisoner's Dilemma agents to evolve within a spatial framework [5, 70] has an impact on both the chance that cooperation will arise and on which strategies evolve. Other version of the Prisoner's Dilemma are also studies with evolutionary computation. In [47], the authors implement multiple levels of cooperation and defection and include noise. However, the choice of representations in many study might partly affect consequences of their experiments. Representation is one of the many parameters which can be tweaked in an evolutionary algorithm to change its behavior. It defines the search space and how the algorithm moves through it.

As previously discussed, in a co-evolutionary algorithm representation it is an even more important parameter because of the effect it has on fitness evaluation. How fit an individual is depends on the composition of the population he is being evaluated with. Strategies whose success depends on exploiting other strategies are only fit when in the company of exploitable strategies. Other strategies can only be successful when there is a threshold number of other strategies like them present in the population [34]. In this section we review results of representational sensitivity study in the IPD using fingerprints to discuss potential causes for the results observed.

2.3.1   EXPERIMENTAL DESIGN

In order to study the sensitivity of the IPD to the choice of representation, a single evolutionary algorithm design based on that used in the relatively early study by Miller [85] was chosen. In this design, a population of agents is evolved using round-robin play of the IPD as a fitness function. A two-thirds elitism is used with the remainder of the population replaced by performing fitness proportional selection with replacement on the elite to generate pairs of parents. The elite are the highest scoring agents. These parents are then copied and subjected to representation specific crossover and mutation operators. All other features of the evolutionary algorithm (population size, number of generations, fitness evaluation) were uniform across all experiments. The data structures used were: finite state Mealy machines with both a direct and a cellular encoding, Boolean formulas with and without a time-delay operation, ISAc lists, Markov chains, look-up tables, two types of feed-forward neural nets one of which is biased toward cooperative behavior, and the Boolean function stacks described in Section 2.1.

The evolutionary algorithms used operate on a population of 36 agents. Agent quality is assessed by a round-robin tournament in which each pair of players engage in 150 rounds of the IPD. Reproduction is elitist with an elite of the 24 highest scoring strategies. When constructing the elite, ties are broken uniformly at random. Twelve pairs of parents are picked by fitness-proportional selection with replacement on the elite. Parents are copied, and the copies are subjected to variation of the sort defined for the representation in question. Variation consists of the representation-specific crossover and then the representation-specific point mutation operator done a number of times selected at random from a Poisson distribution with a mean of 1. All representations permit access to information about three past actions of the opponent with actions before the start of a round of play assumed to have been cooperation.

In each simulation the evolutionary algorithm was run for 250 generations. The number of populations with a mean fitness of at least 2.8 and at least 2.25 were saved at generations 50, 100, 150, 200, and 250. For finite state machines with 16 states, an average fitness of 2.8 represents at most sporadic defection before a pair of machines fall into pure cooperation that continues indefinitely (see [100] for details). In contrast, the threshold of 2.25 is the score that a pair of players playing one another uniformly at random would achieve. Many of the parameters used in this study are either selected in imitation of previous studies or essentially arbitrarily. A careful parameter-variation study would require millions of collections of evolutionary runs because of combinatorial explosion of the parameter space. Despite the arbitrary parameter choices, the study demonstrates profound representational sensitivity in Prisoner's Dilemma playing agents.

2.3.2   EXPERIMENTAL RESULTS

A set of 400 simulations with distinct random initial populations were performed for each representation. At generations 50 and 250, the probabilities were computed of a given simulation having an average fitness in excess of 2.8 or an average fitness in excess of 2.25. In [100], it is shown that scores above 2.8 in 150 rounds of play require that finite state machines with 16 (or fewer) states must be engaging in sustained mutual cooperation after a transient set of moves that may include some defections. In this study 2.8 is taken as a surrogate for essentially cooperative play. The score 2.25 is the average for random play, in which players flip a fair coin to decide if they are cooperating or defecting.

In subsequent figures and tables the representations are abbreviated as follows: neutral neural nets, **NNN**, cooperative neural nets, **CNN**, Boolean formulas with delay operations,

DEL, ISAc lists, **ISC**, look-up tables, **LKT**, Markov chains, **MKV**, Boolean formulas, **TRE**, directly encoded finite state machines, **AUT**, cellularly encoded finite state machines, **CAT**, and function stacks, **FST**. It is evident that there was substantial variation in the level of cooperation as a function of the choice of representation. The status, cooperating or not or better-than-random or not, of each collection of runs was treated as a Bernoulli variable. The normal approximation to the binomial was then used to obtain confidence intervals for the number of cooperative or better-than-random simulations.

Figure 2.8 shows the probability of essentially cooperative play and better than random play for populations using the specified representations in generation 250. Figure 2.9 reports the same statistics, but for generation 50 of evolution rather than generation 250. Notice that the outcomes vary, as the result of changing the agent representation, from pure defection to over 90% sustained cooperative play.

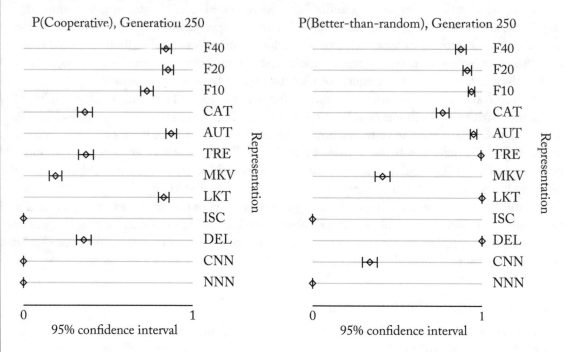

Figure 2.8: Probabilities of essentially cooperative behavior and better-than-random behavior in generation 250 for each of the representations tested.

The probability of a representation displaying essentially cooperative behavior segregated the representations into four groups: (1) ISAc lists and both kinds of neural nets did not display any cooperative behavior; (2) Markov chains were in a group by themselves, displaying a low level of cooperative behavior; (3) Boolean formulas, both with and without the delay operation, together with the cellular encoding of finite state machines displayed an intermediate level of

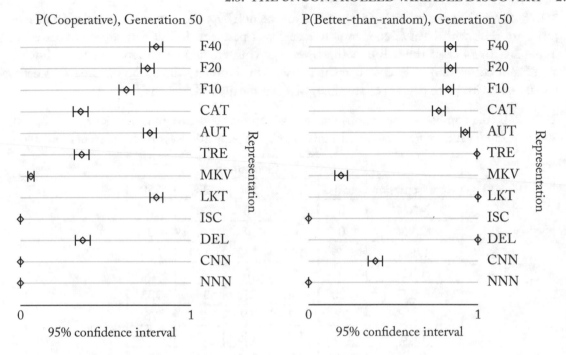

Figure 2.9: Probabilities of essentially cooperative behavior and better-than-random behavior in generation 50 for each of the representations tested.

cooperation below 50%; and (4) The direct encoding of finite state machines, look-up tables, and function stacks displayed a high level of cooperation.

These groupings were evident at both generation 50 and generation 250, though the Markov chains were substantially more cooperative in generation 250 than in generation 50. The probability that a representation would cooperate at a rate better-than-random yields different groupings than those for essentially cooperative play. ISAc lists and neutral neural nets are unable to even play better-than-random. Markov chains and cooperative neural nets attain a moderate rate of better-than-random play, not far below 50%. Function stacks and both encodings of finite state machines achieve a high rate of better-than-random play (though the cellular encoding is significantly lower). It is worth noting here that the data indicate that the initial state of the cellular encoding system, tit-for-tat, did not apparently yield a bias toward cooperation. Both sorts of Boolean formulas and the look-up tables play better-than-random at least 398 times out of 400. These groupings were the same in generations 50 and 250, except that the Markov chains were, again, significantly worse in generation 50.

Given the apparent near-stasis of all representations except Markov chains between generation 50 and 250, the number of essentially cooperative populations at generations 50, 100,

150, 200, and 250 was tabulated and a t-test for the null hypothesis that the correlation of evolutionary time with level of cooperation is zero was performed. The data are summarized in Table 2.4. In all cases the null hypothesis was not rejected, and so no representation shows a significant correlation of level of cooperation with evolutionary time. The maximum statistic, $t = 0 : 463$ with three degrees of freedom, was that for the Markov chains.

Table 2.4: The number of populations that are essentially cooperative in generations 50, 100, 150, 200, and 250 for each representation. The total number of populations for each representation is 400.

Representation	Generation				
	50	100	150	200	250
NNN	0	0	0	0	0
CNN	0	0	0	0	0
DEL	149	143	142	144	144
ISC	0	0	0	0	0
LKT	320	340	330	343	333
MKV	27	43	59	74	76
TRE	147	131	146	138	149
AUT	306	344	351	350	351
CAT	144	155	145	154	146
FST	293	320	327	336	335

Fingerprint Analysis of Complex Representations   Fingerprints permit the automatic filtration of sets of evolved strategies. Table 2.5 tabulates the number of times fingerprints similar to each of the eight strategies fingerprinted in earlier appeared for three of the representations. These data make it clear that the representations are locating strategies with very different frequencies. The choice to use the two finite state representations and function stacks for fingerprint analysis was made because, with the exception of ISAc lists, these representations are the most computationally complex, are relatively cooperative, and include a pair or representations with identical strategy spaces (the two representations of finite state machines).

The criterion for deciding a fingerprint was of a given type is as follows. A grid of 25 points in the interior of the unit square are used to sample the fingerprint. These points are all those of the form $(\frac{i}{6}, \frac{j}{6})$ where $0 \leq i, j \leq 6$. For the reference strategies, AllD, AllC, and so on, the values at these points are determined by using equations in Section 2.2. For the finite state machines and function stacks, the value of the fingerprint is determined by sampling play against $JA(TFT, x, y)$ at the 25 values of the noise parameters. For any particular value of the noise parameters, sets of 150 rounds of the IPD were played repeatedly until the variance of the estimate of the fingerprint value dropped to 0.005.

Table 2.5: A tabulation for both finite state representations, as well as for function stacks, of how often several identifiable strategies appear

	Representation		
Strategy	AUT	CAT	FST
A11D	99	18	1227
A11C	19	18	339
TFT	610	234	1411
PSY	0	10	240
PAV	12	2	7
TF2T	4	5	5
TTFT	1	0	227
Pun1	147	2	4
Other	2719	3311	140

The distance matrix D for the eight reference strategies is given in Figure 2.10. The value 0.05 was chosen for classifying a strategies as similar to a reference strategy because the closest that any two reference strategies approach is 0.17.

$$D = \begin{bmatrix} 0.00 & 1.91 & 1.07 & 1.02 & 0.99 & 1.38 & 0.75 & 1.38 \\ 1.91 & 0.00 & 0.90 & 1.02 & 0.93 & 0.61 & 1.24 & 0.55 \\ 1.07 & 0.90 & 0.00 & 0.59 & 0.32 & 0.32 & 0.36 & 0.35 \\ 1.02 & 1.02 & 0.59 & 0.00 & 0.28 & 0.73 & 0.66 & 0.62 \\ 0.99 & 0.94 & 0.32 & 0.28 & 0.00 & 0.50 & 0.45 & 0.43 \\ 1.38 & 0.61 & 0.32 & 0.73 & 0.51 & 0.00 & 0.68 & 0.17 \\ 0.75 & 1.24 & 0.36 & 0.66 & 0.45 & 0.68 & 0.00 & 0.70 \\ 1.38 & 0.55 & 0.35 & 0.62 & 0.43 & 0.17 & 0.70 & 0.00 \end{bmatrix}$$

Figure 2.10: Distance matrix for reference strategies.

The number of AllD fingerprints found for the function stacks requires explanation. If each of those strategies is in fact AllD, then the level of essentially cooperative behavior exhibited, near 90%, is impossible. Fingerprints are computed from the asymptotic behavior of a player against a noisy version of another strategy (tit-for-tat in this study). An AllD fingerprint thus need not be AllD; rather there is a probability of $1 - \epsilon$, for any small ϵ, that the machine will play AllD against a player when noise is present. In the noise-free Prisoner's Dilemma used as the fitness function in this study, a strategy playing against a copy of itself could use only transient states of the Markov chain used to compute the fingerprint and thus avoid the asymptotic AllD behavior which it exhibits against a noisy opponent. The AllD play is quite likely to be where "strangers" as well as players with noise end up when playing against a strategy with an AllD fingerprint. This, in turn, suggests that the use of transient states is common in function stacks.

Hand examination of the evolved function stacks yielded examples of the vengeful strategy that cooperates until its opponent's first defection and defects thereafter. This is an example of a strategy that is completely cooperative when playing copies of itself but which fingerprints as AllD. This behavior is one that is already known from in finite state machines. A sequence of plays, called a handshake, permits a player to recognize copies of itself or close relatives. This behavior is one that is already known from [30] in finite state machines. A sequence of plays, called a handshake, permits a player to recognize copies of itself or close relatives. Any player not using this sequence of plays drives the machine into a part of its state space rich in defection, often a single state that does nothing but defect. Table 2.5 suggests this behavior is more common among function stacks than finite state machines.

2.3.3 CONCLUSION

The most cooperative representations found in this study were finite state machines in their direct representation, function stacks, and look-up tables. A sequence of five ISAc nodes can simulate a state of the type of finite state machines used in this study and so, computationally, the ISAc list representation included the strategies coded by the finite state machines. In spite of this, the ISAc lists displayed the most divergent possible behavior from directly represented finite state machines. Boolean formulas without the delay operation and both sorts of neural nets encode exactly the 256 Boolean functions on three variables that are also coded by look-up tables. These four representations, encoding precisely the same strategy space, include the single most cooperative and the two least cooperative representations studied. This demonstrates conclusively that the representation and its interaction with selection and variation via the variation operators completely dominates the outcome of this sort of simple experiment.

In a similar but less spectacular fashion the behavior of the two different encodings of finite state machines demonstrate sensitivity to representation. In [76], the completeness of the cellular representation is demonstrated: the cellular representation can generate any finite state machine if enough editing rules are used. The two representations thus encode precisely the same strategy space while exhibiting significantly (and substantially) different levels of essentially cooperative and better-than-random play. The two sorts of Boolean functions, with and without the delay operation, code for substantially different strategy spaces. The delay operations increase the time-horizon of the agents, give them a limited type of internal state information. This combinatorially increase the number of strategies encoded by the delay agents over the plain ones. In spite of this, the two sorts of Boolean formulas are not separated by their essentially cooperative or better-than-random play at generation 50 or 250. In this case the similarity of the agent representation apparently swamps differences between the strategy spaces.

The experiments reviewed here indicate that the interplay of population initialization, variation operators, selection, and reproduction with a particular choice of IPD simulation substantially change the distribution of population types that appear. The question "why?" is left to the future. Finer resolution studies within a single representation type may help to answer this question. The following list of potential causes for the observed variation is diffidently suggested.

First, population initialization produced initial populations with varying levels of cooperative vs. exploitative agents to a degree that changed the eventual distribution in evolved populations. While evolution exhibits an ability to recover from a lack of cooperation, this varies from representation to representation.

Second, the variation operators, crossover and mutation, created agents that were easy to exploit to varying degrees in different representations. Experience suggests that a supply of exploitable agents promotes exploitation and so this is a reason that the exact mechanism of the variation operators may matter critically.

Third, the fidelity of reproduction, the probability a child would resemble its parents, varied across representations. A population in which there is frequent appearance of novel strategies would require a fairly high level of defensive play. Low reproductive fidelity could, thus, drive low levels of cooperation.

Fourth, the granularity of a representation is a measure of the size of the minimum change in behavior that strategies coded in that representation can make via application of the variation operators. The granularity of representations could strongly affect the evolution of cooperation. Cooperation must be reciprocated to be sustained. A representation that can defect "just a little bit" or "occasionally" greases a slippery slope to exploitation and then uniform defection. Finite state machines have very coarse granularity while the Markov chains used in this study have relatively fine granularity (they can continuously vary their probability of defection).

These explanations should not be viewed as mutually exclusive and are almost certainly far from complete. It is possible for all of them to be responsible for some of the observed effect and as yet undiscovered causes for these effects seem certain. Having demonstrated a large impact for representation, we now move on, in the next chapter, to looking at other problematic quantities.

CHAPTER 3

Problems Beyond Representation

This chapter chronicles selected examples from a long series of attempts to document which factor affects the outcomes of evolutionary training of agents playing simple mathematical games. Many recent studies attempt to model human or animal behavior by evolving agents to play a mathematical game, and do so without controlling for factors with a large potential impact on the results. In this chapter we follow up on the results from changing the agent representation in Chapter 2 and examine other factors that can have a significant impact on evolutionary agent training systems.

3.1 CHANGING THE PAYOFF MATRIX CHANGES THE AGENTS THAT ARISE

In the IPD the agents play many rounds of Prisoner's Dilemma with the payoff matrix defined previously. The inequalities defining that matrix permit a broad variety of different payoff matrices. IPD is widely used to model emergent cooperative behaviors in populations of selfishly acting agents and is often used to model systems in biology [99], sociology [69], psychology [96], and economics [67]. In most studies, the choice of a payoff matrix is made without much concern, and sensitivity to the choice of payoff matrix is not studied. Such a comparison requires care: when a different payoff matrix is used, it is obvious that the scores made by players will change. A score-neutral method of assessing the agents is required. In this section we use fingerprinting introduced in Section 2.2 for this purpose.

Other researchers have considered the effect of changing the payoff function in various forms of the Prisoner's Dilemma. In [98], the authors used a look-up table representation for the N-player IPD using payoff functions that varied over the course of play. This result is one of the earliest, though not directly comparable to the current study. Self adapting payoff matrices were studies in [46] using a very simple depth one memory representation for the agents. In [95], the authors study the effect of time-varying payoffs on a spatially structured population of depth zero agents (cooperators and defectors). While all three of these studies do investigate the effect of varying payoffs, they do so within each simulation. In this section, we investigate the impact of varying a fixed payoff matrix between simulations.

3.1.1   DESIGN OF EXPERIMENTS

The primary purpose of this study is to compare agent populations evolved with different payoff matrices. In order to have a notion of which payoff matrices are extreme we adopt the normalization appearing in [74]. For a collection of payoff values S, D, C, T we apply the transform

$$f(p) = \frac{p - S}{C - S} \tag{3.1}$$

which forces $S = 0, 0 \leq D \leq 1, C = 1$, and $1 \leq T \leq 2$, the latter as a consequence of the original Prisoner's Dilemma inequalities. This transformation maps all payoff matrices into the unit square given in Figure 3.1.

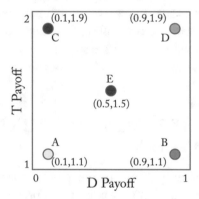

Figure 3.1: The space of normalized payoff matrices showing the position of the five matrices used in this study. The coordinates of this space are the values of the D and T payoffs. The other payoffs have the fixed values $S = 0$ and $C = 1$. Colors used are constant throughout the section.

For use in this section, five payoff matrices were selected. Four of these matrices are placed within payoff matrix space at a position 10% short of the four corners. It is intuitive that placements on the boundary of the space might yield pathological behaviors; it is our desire to compare payoff matrices that might appear in practice. The fifth payoff matrix is placed in the center of the normalized space.

Agent Representation   The agent representation used in this section is 8-state finite state machines with actions associated with transitions between states (Mealy machines). Finite state machines were chosen as the agent representation in this study because they are capable of encoding a broad variety of Prisoner's Dilemma strategies [21] and are easy to fingerprint [19]. State transitions within the finite state machines are driven by the opponent's last action. Access to state information permits the machine to condition its play on several of its opponent's previous moves. The machines are stored as linear chromosomes listing the states. The initial state and action for a given machine are stored with and undergo crossover with the first state in this linear chromosome.

Two variation operators are employed, a binary variation operator and a unary variation operator. The binary variation operator used is two-point crossover on the list of states. Crossover treats states as atomic objects. The unary variation operator (mutation) changes a single-state transition 40% of the time, the initial state used by the machine 5% of the time, the initial action 5% of the time, or an action associated with a transition 50% of the time. Mutation replaces the current value of whatever it is changing with a valid value selected uniformly at random.

The Evolutionary Algorithm   The evolutionary algorithm used here operates on a population of 36 agents, a number chosen for compatibility with previous studies [19, 20, 21, 26]. Agent fitness is assessed with a round-robin tournament in which each pair of players engage in 150 rounds of the IPD. Five experiments were performed, one with each of the payoff matrices given in Figure 3.1, consisting of 30 replicates. The experiments are called Experiment A, B, C, D, and E based on the labels for the payoff matrices given in Figure 3.1.

Reproduction within the algorithm is elitist with an elite of the 24 highest scoring strategies, another choice that maintains consistency with past studies. When constructing the elite, ties are broken uniformly at random. Twelve pairs of parents are picked by fitness-proportional selection with replacement on the elite. Parents are copied, and the copies are subjected to crossover and mutation.

In each simulation the evolutionary algorithm was run for 6,400 generations. The elite portion of the population in generations 50, 100, 200, 400, 800, 1,600, 3,200, and 6,400 was saved for analysis. Each of these distinct sample times are called *epochs*. This yields 30 sets of 24 machines at each of 8 epochs for each of the 5 payoff matrices. A number of descriptive statistics are saved for each generation of each replicate. These include the mean fitness, a 95% confidence interval on the fitness, and the maximum fitness.

Many of the parameters used in this study are either selected in imitation of previous studies or are essentially arbitrary. A careful parameter-variation study would require millions of collections of evolutionary runs because of the combinatorial explosion of the parameter space. The current study represents a small slice of this diverse space of possible experiments. We now carefully describe the ways in which the evolved machines were analyzed.

Analysis Techniques Used   The most commonly applied measure to determine if a change in the setup of an evolutionary system for training Prisoner's Dilemma agents changes its behavior is the probability that a given population is cooperative. The past studies [11, 15, 19, 26, 30], all of which used the payoff values $S = 0$, $D = 1$, $C = 3$, $T = 5$, established that, when 150 rounds of IPD are used in fitness evaluation, an average score of 2.8 signifies that machines end up in a cycle of sustained cooperation. Trying to compute the destination of this critical value via the affine normalization is difficult—the exact mix of interactions is not preserved by the normalization. To avoid this problem, the saved populations were run through fitness evaluation with the 0,1,3,5 payoff values to assess if they were playing in a cooperative manner or not. This gives a measure of "probability of cooperation" that matches the one used in the earlier studies.

In [19], a technique for binning evolved agents was employed. Strategies that can be implemented with finite state machines using no more than two states have twelve distinct fingerprints. Table 3.1 list 12 strategies that possess these fingerprints. We call these strategies *bin exemplars*. In order to bin strategies, the sampled fingerprint of the strategy under test is compared to the sampled fingerprints of the 12 bin exemplars. A strategy is placed in the bin corresponding to the bin exemplar with a fingerprint at smallest Euclidean distance from its own. The counts in these bins form a twelve-parameter description of the distribution of strategies.

Table 3.1: Strategies used for binning evolved agents. Note initial actions are not given because initial actions do not affect the fingerprint.

Abbreviation	Name	Description
ALLD	Always defect	This strategy always defects.
2TFT	Two-tits-for-tat	This strategy defects twice in response to defection and otherwise cooperates.
TFT	Tit-for-tat	This strategy does whatever its opponent did last time.
TF2T	Tit-for-two-tats	This strategy cooperates except after a sequence of two defections.
UC	Usually cooperate	This strategy cooperates except after a C following a D.
ALLC	Always cooperate	This strategy always cooperates.
(2TFT)	Inverse two-tits-for-tat	This strategy does the opposite of what 2TFT would do.
PSYC	Psycho	This strategy does the opposite of what its opponent did last time.
TFT-PSY	Tit-for-tat-psycho	This strategy plays like TFT until its opponent defects; then it plays like PSYCHO until its opponent cooperates.
PSY-TFT	Psycho-tit-for-tat	This strategy plays like PSYCHO until its opponent defects; then it plays like TFT until its opponent co-operates.
(TF2T)	Inverse tit-for-two-tats	This strategy does the opposite of what TF2T would do.
UD	Usually defect	This strategy defects except after a D following a C.

We also apply two techniques from [19] for evaluating the competitive ability of agents. The first, *competitiveness*, matches pairs of agents evolved in different manners. Each pair is treated as a Bernoulli trial and the probability that agents from one group will beat those from another are estimated with 1,000 pairwise trials. The second technique for assessing competitive ability is *relative tournament ranking*. This technique creates Prisoner's Dilemma tournaments with agents drawn from different experiments (and no two from the same evolutionary replicate within an experiment). The mean tournament rank is estimated by sampling 1,000 tournaments. This second assessment is also a measure of competitive ability but a different one. It is possible for an agent to lose in direct confrontation with another agent, but to beat it in a tournament. Both these measures use the $S = 0$, $D = 1$, $C = 3$, $T = 5$ payoff matrix common in earlier studies to provide a common currency for comparison. After normalization this matrix lands in the C-E quadrant of the space shown in Figure 3.1, a little about the line joining C and E and slightly closer to E.

3.1.2   RESULTS AND DISCUSSION

The probability of a population being cooperative is summarized for all experiments and epochs in Figure 3.2. This is the oldest of the analytical tools used by our group for distinguishing different methods of evolving Prisoner's Dilemma agents. In this case it demonstrates that cooperativeness clearly places experiments A, C, E in one group and experiments B, D in another. The two relatively uncooperative experiments are those with high D payoffs, or alternatively, with D payoffs close to the C payoff. This strongly suggests that having the C and D payoffs close together is the enemy of cooperation. Experiment A is also shown to be the one most likely to yield a cooperative population with all 30 replicates becoming completely cooperative in the final epoch. Experiment A exhibits extreme behavior in the other analyzes.

The most complex analysis of the data performed is the binning of strategies according to their Euclidean fingerprint proximity to the 12 reference strategies appearing in Table 3.1. This binning of the strategies from each experiment is shown in Figure 3.3. The overall story told by this analysis is that there is a huge effect on strategy density that results from changing the payoff matrix. A χ^2-test, performed for each epoch, found that the distribution of strategies for different payoff matrices was significantly different with $p = 0.001$. There is also a smaller effect on strategy distribution resulting from permitting evolution to continue from one epoch to another. The ACE/BD division of the experiments remains apparent in this analysis, but no two payoff matrices exhibit the same behavior for all bins.

We note that experiment A has a higher weight in the bins whose exemplars are nice (TFT, UC, ALLC, TF2T), while noting that the latter three of these exemplars are also exploitable. Experiment A has a low D payoff and a low T payoff so that the relative worth of pure cooperation is higher in experiment A than in any other. The data suggest that defection arises at such a decreased rate that even exploitable strategies are not challenged too often.

Strategies in the AllD and TFT bins are well represented in all experiments and epochs, except that the TFT bin becomes much less significant in the later epochs within experiment B while AllD gains. Somewhat counter-intuitively, the TF2T bin, whose exemplar is both nice

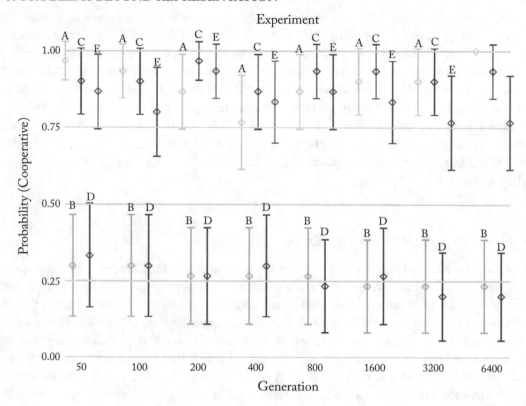

Figure 3.2: Shown are the probabilities that a population, at a given epoch and in a particular experiment, will be cooperative. These confidence intervals are computed with a normal approximation to the binomial distribution and so yield probabilities of cooperation above 1.0 when the probability of cooperation is relatively high. Such results should be viewed as truncating at 1.0.

and exploitable, gains sharply in experiment B in the last epoch. It is likely that this increase is among non-exploitable strategies within the bin.

The three bins whose exemplars are both not nice and exploitable, (2TFT), Psycho, and (2TFT), are poorly represented, but some effective strategies in the (TF2T) bin arise for Experiments A, C, and E in the later epochs. It is important to remember that there are millions of strategies in each of these bins and some of them can be far more competitive than their exemplars.

Figure 3.4 shows the results of placing pairs of individuals from different experiments but the same epoch into competition. For each pairing in this analysis, a normal approximation to the binomial was used to test the null hypothesis that the probability of agents from one of the two experiment had a probability not significantly different from one-half of beating agents

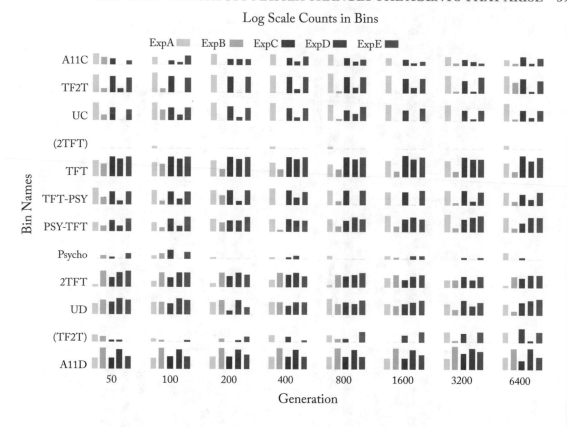

Figure 3.3: Logarithmic strategy density for all eight epochs for all five experiments with strategies binned by fingerprint proximity to the strategy naming the bin. These strategies are described in Table 3.1.

from the other experiment. In 79 of 80 comparisons this null hypothesis was rejected. Only experiments A and C fail to demonstrate a competitive advantage one way or the other, and they only do so in the third epoch (generation 200).

The hypothesis that experiment A produces populations highly enriched with nice, exploitable strategies is supported by this analysis. The A vs. B and A vs. D comparisons show that the individuals chosen from experiment A are crushed in all epochs. In these individual match-ups, the agents drawn from experiment B have an advantage against all opponents in all epochs. This seems to simply be the result of agents from experiment B being quite obnoxious (defecting a lot). On the other hand, Figure 3.2 tells us that some fraction of the agents from experiment B can cooperate within their own populations. This, in turn, makes it likely that the agents in B have some sort of self-recognition technique or *handshake* [19]. A handshake is a sequence of a few initial moves that an agent uses to recognize close kin or copies of itself.

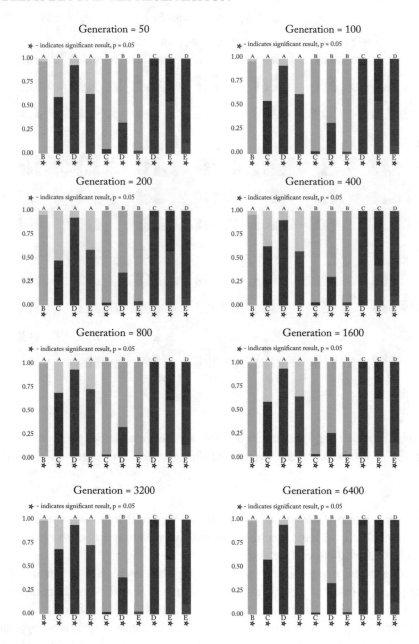

Figure 3.4: Pairwise competitive advantage, agent vs. agent, for all pairs of experiments over all sampled. Fraction of victories in 1,000 samples shown by color. Results are statistically significant if a 95% confidence interval on the binomial parameter fails to include $p = 0.5$.

Experiment D seems to be experiment B's understudy. In all epochs it soundly beats all opponents except B. Agents from D do better against agents from B than do the agents from any other experiment. Experiments C and E are the only two that do not have the same ordering of competitive advantage across all epochs. They are also the most similar to one another in Figure 3.3.

Figure 3.5 gives the results on average tournament rank for agents from all five experiments. Each tournament had six agents from each experiment, drawn from the same epoch but distinct replicates. This latter constraint was intended to remove genetically based collaboration from the list of confounding factors. This analysis showed the most epochal variation with experiments B, C, and D, making statistically significant exchanges in their relative position in the rankings in different epochs. The majority of comparisons made possible by this experiment are statistically significant.

The performance of agents from experiment A confirms the earlier indications that these agents are both nice and exploitable. The average tournament ranking of agents from experiment A is the worse in the first epoch and goes down from there. It is worth noting that the presence of the milksop agents from experiment A probably boosted the tournament rank of the aggressive agents from experiments B and D. This makes the dominance of the relative tournament rankings by agents from experiment E, with a payoff matrix in the center of the normalized payoff matrix space, more startling.

Figure 3.4 suggests that A, C, and E manage to cooperate quite a lot within populations, but that some agents from C and E exploit agents from A at least part of the time. This, in turn, suggests that the abysmal average ranking of agents from experiment A are largely the result of exploitation by agents from B and D. The dominance of experiment E suggests that these tournaments are recapitulating Axelrod's tournament results in which TFT, a nice agent that could defend itself, won both tournaments.

The payoff matrix used in experiment C rewards behavior that generates T payoffs but discourages behavior that results in sustained defection. The simplest version of this encouraged behavior is not sustainable under the action of evolution, suggesting that agents drawn from experiment C will have a higher fraction of agents that try to get away with something, i.e., that constantly test their opponents but "apologize" for reciprocated defections. This behavior, apparently, serves them poorly in competition with agents from experiments D, E, and especially B.

Generation 200, the second epoch, shows signs of being a transition from one behavior to another. Experiment A does better in this epoch than it does in any other while C does worse. Experiments B and D cross between this epoch and the previous one. Some large changes (UC, Psycho) are visible in Figure 3.3 between this epoch and the one before it. This is the epoch in which experiments A and C "tie" for individual competitiveness. The average tournament ranking analysis demonstrates that the character of the populations do change over time, although this difference is smaller than that caused by changing the payoff matrix.

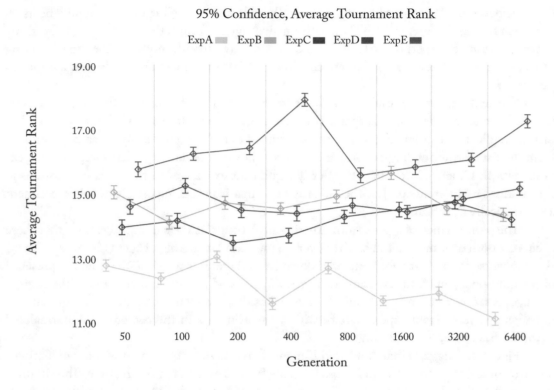

Figure 3.5: Average tournament rank over $n = 1,000$ sampled 30-player tournaments with 6 participants from each experiment. Participants in a tournament are never drawn from the same replicate within an experiment. Results are calculated for all epochs and members of a given tournament are drawn from the same epoch.

3.1.3   CONCLUSIONS

We demonstrate conclusively that changing the payoff matrix, within the bounds permitted by the defining inequalities of Prisoner's Dilemma, yields different results. The distribution of strategies present, the probability of evolving a cooperative population, the competitiveness of individual agents and competitiveness in a tournament all change significantly.

The largest effect, separating experiments A, C, and E from experiments B and D, comes from having C and D payoffs that are close together. This makes the slide into mutual defection less expensive. This effect is stronger in experiment B, where the T payoff is also close to the D payoff.

In [19], agents with two different sorts of competitive ability *contest winners* and *evolution dominators* were defined. A contest winner will receive a high tournament ranking when playing round-robin against a slate of diverse opponents. An evolution dominator will tend to cooperate with agents very similar to itself (its relatives) and behave aggressively against agents different from itself. These definitions reflect the different objective functions used for scoring contests and for driving the evolution of a small well-mixed population.

The experiments here are all of the sort that generate evolution dominators: small well-mixed populations of close relatives. The degree to which a strategy is an evolution dominator is strongly influenced by the choice of payoff matrix.

Remembering that Figure 3.3 uses a logarithmic scale, it is clear that the analysis by binning alone shows that all five payoff matrices generate distinct mixes of agents. The two experiments that seem the most similar in this analysis, C and E, are separated in every epoch by both the individual competitiveness and tournament ranking analysis.

Application of the results presented here will depend on the goals of the researcher applying them. The results suggest the following rules of thumb for designing a payoff matrix for a specific purpose.

1. Having the cooperate and defect payoffs close together yields aggressive agents, likely to cooperate only with agents similar to themselves or, possibly, with no one.

2. It is possible, by having a low D payoff and a low T payoff, to generate milksop agent that are highly cooperative but bad at defending themselves from aggression.

3. Placing a different slant on rule 2, optimizing the payoff matrix to generate cooperation can yield agents with very low competitive and defensive ability.

4. The value of the defect payoff has more influence on the trajectory of evolution than the value of the temptation payoff. This is probably because it is used more often—strategies that lead to getting temptation payoffs are typically not sustainable in evolution because their victims fall prey to selection.

5. If we accept that contest winners are generalists, intermediate values in the payoff matrix, as exemplified by experiment E are better for yielding generalist strategies.

6. The duration of evolution has some effect on the impact of the choice of a payoff matrix. This effect is smaller than the effect of changing the payoff matrix.

Next Steps   The results in this study treat a single, if much used, evolutionary algorithm for training Prisoner's Dilemma agents. Our main result is that this algorithm is highly sensitive to the choice of payoff matrix. A very interesting direction would be to see if other evolutionary algorithms or agent representations are less (or more) sensitive to the choice of payoff matrix. In [35], it was shown that using a spatially structured evolutionary algorithm yielded a population that was both cooperative and stable against invasion. Use of a similar spatial structure might reduce sensitivity to choice of payoff values. It also seems likely, given the results in [21], that changing the agent representation will change the sensitivity to choice of payoff matrix but the probable direction of the change is not at all obvious.

In [19], it was found that noise had a substantial effect on which strategies arose during evolution and on the competitive abilities of those strategies. Many of the analysis techniques used here were drawn from this paper. This study checks for a sensitivity that is orthogonal to sensitivity to noise. This suggests that varying noise level and payoff values might generate substantial effect whose exact nature is not obvious.

Several other analyzes we not performed on the machines evolved in this study. They include: evolutionary velocity [19], relative complexity of strategies as exemplified by the number of states in the reduced finite state diagram, and time-of-appearance of particular strategies found to appear only in later epochs [37]. This study could be extended to measure other effects of changing the payoff values.

Finally, we note that Prisoner's Dilemma represents a unit square in the infinite space of normalized payoff matrices. Many other interesting games, such as the hawk-dove game, the graduate school game, and others, can be checked for the sort of sensitivity demonstrated in this study.

3.2   CHANGING THE LEVEL OF RESOURCES

This section follows and refines the material presented in Section 2.3 on changing the agent representation. The results in this section demonstrate that, even within a single agent representation, changing the level of resources available can have a large impact on the results. It may be that the boundaries between representations were drawn in an arbitrary way, one that arises from the way we choose to code agents, and changing the resources available to a type of agent might mean that you are using a different representation. Studying the impact of changing resource levels avoids needing to settle that philosophical question.

The resources to be varied come in two forms—computational resources like states in a finite state machine or neurons in an artificial neural net and resources such as the quantity of information available about an opponent's past actions. The twin hypotheses that propose that changing computational or informational resources changes the type of agents that evolve are examined in this section.

This work is motivated by a study investigating if zero-sum games were more likely to evolve intransitive collections of agents [7], summarized in Section 4.1. In this study it was found that changing the number of states in a finite state representation had a large impact on the type of agents that evolved. Changing the number of nodes or states available to agents are examples of changing a computational resource. One of the two hypotheses tested in this section is that changing the level of computational resources available to an agent training system can significantly change the resulting behavior.

A second sort of resource can also modify agent behavior significantly: the information presented to the agent on which to base its decision. We test the second hypothesis changing the type of information available to an agent training system can significantly change the resulting behavior here.

The hypothesis that different representations yield different behavior has already been documented in the earlier studies. The two hypotheses under test in this study are investigating the degree to which sensitivity to the details of the implementation within a representation still yield substantial changes in the behavior of evolved agents. We are comparing finite state machines with more and fewer states, neural nets with more and fewer neurons, Markov chains with more and fewer probability levels available, and look-up table with large and smaller, as well as different, time histories. The between representation comparisons are well settled in the literature. We address within representation comparisons in this section.

3.2.1   DESIGN OF EXPERIMENTS

The evolutionary algorithm used in this section operates on a population of 36 agents, a number chosen for compatibility with Section 2.3. Agent fitness is assessed by a round-robin tournament in which each pair of players engage in 150 rounds of the IPD. Reproduction is elitist with an elite of the highest scoring two-thirds of strategies, also maintaining consistence with Section 2.3. Ties that occur while selecting the elite are broken uniformly at random. Pairs of parents are picked by fitness-proportional selection with replacement on the elite to replace all non-elite members of the population. Parents are copied and the copies are subjected to crossover and mutation. Details are given with the descriptions of the individual representation.

In each simulation the evolutionary algorithm was run for 3,200 generations using 100 replicates (experiments with distinct random number seeds). The elite portion of the population in generations 50, 100, 200, 400, 800, 1,600, and 3,200 was saved for analysis. This yields 100 sets of 24 agents at each of seven epochs. A number of descriptive statistics are saved in each generation of each replicate. This includes the mean fitness and the variance of fitness. The *total score statistic*, described subsequently, uses this saved information.

Some of the parameters used here are selected in imitation of previous studies to enable comparison while others, such as the fairly low mutation rate, are essentially arbitrarily. The resource parameters serving as the focus are varied systematically. A complete factorial design, including all potentially relevant parameters, would require millions of evolutionary runs and so the experiments performed represent a somewhat arbitrary selection guided by intuition and experience.

Agent Representations   Four agent representations are used in this study: look-up tables (LUTs), Markov chains (MKVs), artificial neural nets (ANNs), and finite state machines (FSMs) using the Mealy architecture [73]. These representations were selected from those used in [21] because they include representations that are and are not highly cooperative, a representation that uses random numbers and ones that do not, and representations that have a wide variety of resources that can be varied. These include states in the FSMs, number of neurons in the ANNs, and number of available probability levels for the MKVs. All of the representations, except the FSMs, can have the amount and type of information they are given varied. This selection on representations permits a broad test of both the hypothesis that varying computational resources changes agent behavior and that changing available information changes agent behavior.

It is worth noting that LUTs and ANNs encode exactly the same strategy space in different ways. That strategy space is contained in the space encoded by the MKVs and also in the strategy space encoded by FSMs, but in different ways. The FSM and MKV representations have a substantial disjunction. One of the results from Section 2.3 is that representations that encode the same strategy space do not yield the same results when used as the agent representation in an evolutionary algorithm. Similarly, while a representation like a MKV can encode a LKT it never does. The probability that a MKV will simulate an LKT is very small when the number of probability levels is small, and vanishingly small when the number of probability levels is large.

Throughout this experiment, when an agent bases its behavior on an opponent's or its own past action, actions before the start of play are presumed to be cooperates. This is an arbitrary choice, but one that avoids the complexity of encoding in each agent a set of presumed pre-play actions. While this arbitrary choice may affect agent behavior, it can only reduce variability of behavior, and so does not change the conclusions of [21] which document high levels of variability in the behavior of different types of agents.

Look-up Tables   A look-up table takes some number of past actions by the agent's opponent or the agent themself, transforms them into an index number, and looks up the resulting action. An example of a look-up table, based on the opponents last three actions, is shown in Figure 2.4. LUTs are represented as a string of responses, of length 2^n where n is the number of past actions used to index responses. The instance of the representation for the example given in Figure 2.4 is **CDCDCDCD**. The resource varied for look-up tables is the number and type of past actions used to generate the index. The experiments performed are: 2:2, 3:3, 4:4, 4:0, 6:0, and 8:0 where **a:b** means **a** of the opponents past actions and **b** of your own. The variation operators used are two-point crossover of the string of responses and a mutation operator that changes one entry of the response string.

Markov Chains   A Markov chain is very similar to a look-up table. The difference lies in the encoding of the response which is a probability of cooperation rather than a simple **C** or **D**. The array of probabilities of cooperation form the representation for this agent type. An example of a Markov chain dependent on three inputs is shown in Figure 3.6.

Notice that all probabilities in Figure 3.6 are multiples of one-third. This is an agent with four probability levels. We vary two sorts of resources for the Markov chains: number of actions used for indexing and number of probability levels available. Using the notation for LUTs, we test 3:3 and 6:0 input values. The number of probability levels allowed is one of 4, 32, or 128. These levels are equally spaced and always begin with zero and end with one. Variation operators are two-point crossover of the array of cooperation probabilities and a mutation operator that changes one entry of the response string to a new value chosen uniformly at random from those allowed.

Opponent Last Three Actions	Response
CCC	0
CCD	0
CDC	2/3
CDD	1/3
DCC	1/3
DCD	0
DDC	1
DDD	1/3
Chromosome: 0, 0, ⅔, ⅓, ⅓, 0, 1, ⅓	

Figure 3.6: A Markov chain with three inputs and four available probability levels. Only the response column specifying the probability of cooperation, displayed above as the chromosome, is evolved.

Artificial Neural Nets   The ANNs used in this study have inputs in the form of the opponent's and agent's last actions, as with LUTs and MKVs. Inputs are encoded numerically as $D = -1$ and $C = 1$. Initial connection weights are chosen randomly in the range [-1,1] and neurons used the transfer function given in Equation (3.2)

$$F(t) = \begin{cases} -1 & t < 0 \\ 1 & t \geq 0 . \end{cases} \tag{3.2}$$

An example of an ANN of the sort used here is given in Figure 3.7.

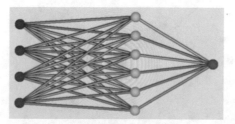

Figure 3.7: This is the structure of an ANN with four inputs, six hidden neurons, and an output neuron. The ANN uses a step function for transition.

The choice of a a step function is made for consistency with earlier studies [4, 16, 21]. An issue that is not examined is the impact that the type of transfer function might have on neural Prisoner's Dilemma playing agents. This is an interesting area for potential future study.

The representation for the ANNs consists of an array of connection weights. The connection weights for each of the input/hidden neuron connections appear in order, relative to the index number of the input, followed by the connection weights from the hidden neurons to the output neuron. Six sets of experiments are performed with ANNs each of which uses either 3:3 or 6:0 inputs. The resources varied is then number of hidden neurons which are set to 3, 9, and 15. Variation operators consist of two point crossover of the array of connection weights and a mutation operator that replaces one connection weight with another selected uniformly at random in the range $[-1, 1]$.

Finite State Machines   The finite state machines used in this experiment use the Mealy architecture with responses encoded on the transitions (the Moore architecture encodes responses on the states). A example of the type of FSMs used in this study appear in Figure 3.8. A finite state machine is always driven by the opponent's last action. For this reason the resource we vary is the number of states available to the machine. In [7] it was discovered that the behavior of finite state machines playing a collection of different 3-move games changed substantially between 8 and 80 state machines, except when the game being played was zero-sum. Since Prisoner's Dilemma is not zero sum, we study the impact of varying the resource "number of states" for FSM agents in the IPD.

The number of states used are one of 4, 8, 16, 32, 64, or 128. The variation operators used are two-point crossovers of the array of states, with the choice of initial state and action attached to the first state, and a mutation operator that changes the machine's initial state or action with a probability of 5% each, changes a transition 40% of the time, and a response 50% of the time. This mutation operator is retained from Section 2.3 and previous studies for consistency.

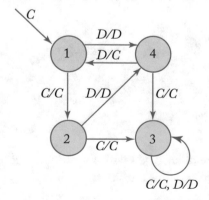

Figure 3.8: A FSM with four states. Transitions are of the form opponents last action/response. The sourceless arrow labeled with C denotes the initial state (1) and action (cooperate).

Analysis Techniques   Two analysis techniques are retained from the preliminary study [16]: *play profiles* and *total score*. We also use competitiveness analysis and nonlinear projection which appear in earlier papers analyzing IPD playing agents. The scoring scheme used in this study, $S = 0$, $D = 1$, $C = 3$, and $T = 5$, ensures that individual scores are in the interval $[0, 5]$ while a population average score must remain in the range $[1, 3]$. We map these scores onto the coloring scheme shown in Figure 3.9.

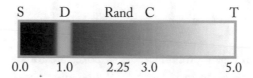

Figure 3.9: The score to color mapping used in figures. The scores corresponding to both defect equilibrium at 1.0, the score expected for to players using coins to play at 2.25, and the score for mutual cooperation 3.0 are picked out with the colors green, red, and blue, respectively.

Play Profiles   One of the primary assessments of an evolutionary system for training Prisoner's Dilemma agents changes is the probability that a given population is cooperative. In past studies [15, 19, 30], it was established that when 150 rounds of IPD with the 0, 1, 3, 5 payoff scheme are used in fitness evaluation that an average score of 2.8 signifies that FSMs with 16 of fewer states end up in a cycle of sustained cooperation. We extend this assessment in this study. We divide the range of possible average scores for an entire population into ten equal intervals, the top one corresponding to the definition of cooperative behavior from earlier studies. At each epoch (50, 100, 200, 400, 800, 1,600, and 3,200 generations) we record the number of populations in each of the ten intervals. The resulting 10×7 table is the *play profile* for an experiment. This method for documenting the character of an evolving population originates in the study extended here. An example of a play profile is shown in Figure 3.10.

Total Score   For each experiment, the average fitness in each generation of each replicate was saved in each generation. A good surrogate for the degree of cooperativeness of agents is the area under these mean fitness curves. An example of this is shown in Figure 3.11. This measure has the advantage of incorporating information from all generations, not just the sampled epochs. Representations that end cooperative, but take a while to get there, will have lower total scores than ones that learn to cooperate immediately. We perform the natural affine normalization to make the minimum area zero (by treating a average score of one as making no contribution) and maximum area one (by treating a mean score of three as making a contribution of $1/G$ where G is the number of generations). The normalized total score for a run of the evolutionary algorithm becomes the reporting statistic.

The total score statistic was developed to give a scalar summary of the type of complex average fitness track shown in Figure 3.11. This statistic was used to assess optimization perfor-

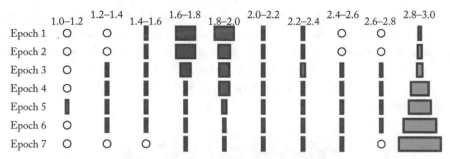

Figure 3.10: An example of a play profile. Colors of boxes are for the score in the middle of the range according to the scheme given in Figure 3.9. The possible range of scores is divided into ten equal intervals, given as top labels, and assessed at each to seven, exponentially spaced, evolutionary epochs. The width of the bars is proportional to the fraction of 100 independent populations in the bar's score range.

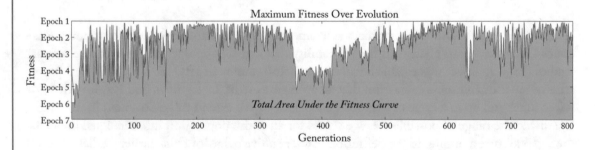

Figure 3.11: The green area is the total fitness under the population average score curve of an evolutionary run. Normalized to have a maximum value of 1, this area is used as a summary reporting statistic for the behavior of a population.

mance in [12, 13]. This is its first use in a co-evolutionary system where fitness is determined by competition. Since an agent population can jump in and out of a cooperative state multiple times, it is difficult to compare different runs. The total score statistic washes out information about *when* cooperation happens, making comparison of different runs simpler. This suggests the statistic is more useful in assessing co-evolutionary training of game-playing agents that assessing optimization.

Competitive Analysis   The competitiveness test used chooses a large number of pairs of agents of different types and then estimates the probability that an agent of one type will beat an agent of another type. Agents with different resource levels can be compared in this manner. This estimate is performed by recording the number of pairs in which a decisive victory (both agents cooperating consistently yields a tie) occurs and then treating victories by one agent type as a Bernoulli variable. An agent is defined to have won a contest if it has the higher average score after 150 rounds of play by a margin of at least a value of 0.05.

State Usage   For the finite state representation there are highly effective representations that can be encoded in a small number of states. The famous strategy *tit-for-tat* can be encoded in one state and the highly robust *fortress-3 and 4* strategies [15] can be encoded with three and four states. Since state transitions are produced initially at random and subsequently by selection-filtered random events there is no requirement that a finite state machines use all of their states. There are two senses in which a state is "used." A state inaccessible by any sequence of transitions from the initial state is strongly unused while a state that is not used against a given opponent is unused against that opponent. The former notion is more stable and easier to pin down. For the finite state, agents will use the fraction of available states used as an assessment.

Nonlinear Projection   *Nonlinear projection* (NLP) [29] is a visualization technique used on high-dimensional data, such as the play profiles used in this study. It is an evolutionary form of non-dimensional multi-metric scaling [78], also called *multi-dimensional scaling* [63]. The goal of NLP is to provide a projection of points from a high-dimensional space into a two-dimensional space that distorts the inter-point distances as little as possible. The projection forms a visualization of the higher dimensional data set. In this study we minimize the *Pearson correlation*, given in Equation (3.3), of the Euclidean distance between play profiles for FSM agents in $\mathbb{R}^{10}$ with the planar Euclidean distance of the projected points. Since this correlation is scale-free it reduces the complexity of the evolutionary search. The resulting projection does not, itself, have a scale; rather, it depicts only relative distances between points:

$$cor = \frac{\sum_{i=1}^{n}(x_i - \overline{x})(y_i - \overline{y})}{(n-1)s_x s_y}, \tag{3.3}$$

where for $z \in \{x, y\}$, $\overline{z}$ denotes the sample mean and s_z denotes the sample standard deviations.

The evolutionary algorithm used to perform NLP uses a population of ten tentative projections stored as lists of points (x, y). The points are initially generated to lie within the unit square with corners $(0, 0)$ and $(1, 1)$. The model of evolution is tournament selection of size seven. Variation operators are two-point crossovers of the lists of points (points are treated as atomic objects that cannot be split by crossover) and two mutation operators, each used 50% of the time. The first mutation operator randomly replaces a point with a new point selected uniformly at random within the unit square. The second adds a Gaussian random variable with a standard deviation of 0.1 to both coordinates of a point. From 1–3 mutations, with the number selected uniformly at random, are performed on any new structure. The algorithm is run for 40,000 mating events, a number chosen by looking at when the fitness for a small initial set of runs leveled off.

3.2.2   RESULTS AND DISCUSSION

Figures 3.12, 3.13, 3.15, and 3.18 give the play profiles for look-up tables, Markov chains, finite state machines, and ANNs. The most cooperative representation, overall are look-up tables. It is not clear which is the least cooperative because there are very different patterns of change in the play profiles over evolutionary time. The most stable representation—with the least change over time—is the ANNs. Look-up tables are equivalent to Markov chains with two probability levels. The Markov chains with four probability levels are as cooperative as the look-up tables with comparable input structures. The play profiles demonstrate that the distribution of levels of cooperation are very different for different representations and change over time in different manners.

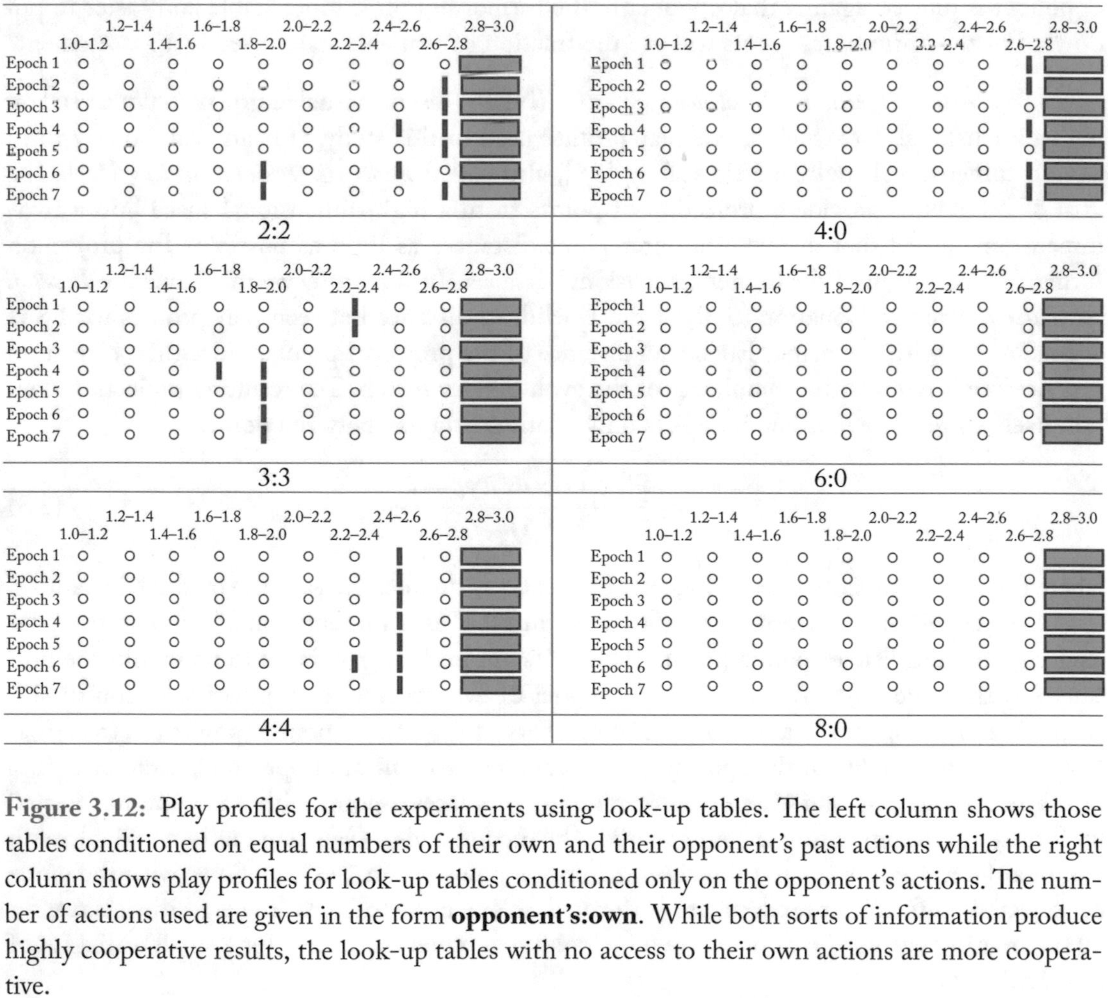

Figure 3.12: Play profiles for the experiments using look-up tables. The left column shows those tables conditioned on equal numbers of their own and their opponent's past actions while the right column shows play profiles for look-up tables conditioned only on the opponent's actions. The number of actions used are given in the form **opponent's:own**. While both sorts of information produce highly cooperative results, the look-up tables with no access to their own actions are more cooperative.

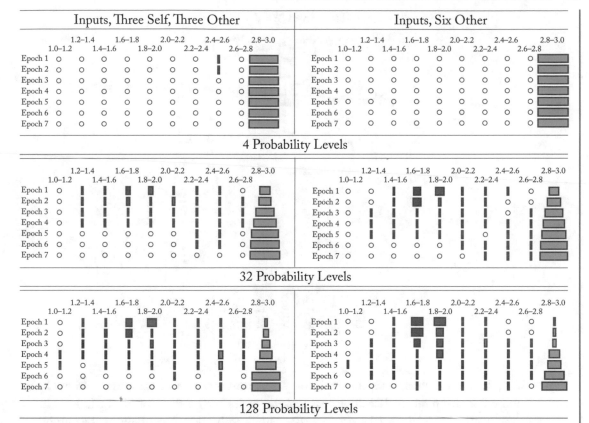

Figure 3.13: Play profiles for the experiments using Markov chains. The left column shows Markov chains conditioned on three of their opponent's actions and three of their own which the right column shows results when the Markov chain is conditioned on the opponent's last six actions. The rows of the table are based on the number of distinct probability levels available to the Markov chain. The most noticeable effect is that increasing the number of probability levels available impairs the emergence of cooperation.

While look-up tables are highly cooperative, the play profiles have a clear pattern. Dividing the inputs between self and opponent yields slightly lower levels of cooperation and using fewer inputs yields lower levels of cooperation. It is also worth noting that Markov chains with four probability levels have play profiles very similar to look-up tables.

Markov chains exhibit the most variable behavior both based on changing the number of probability levels available and changing over evolutionary time. The initial study showed that Markov chains were becoming more cooperative through the fifth epoch. In this study,

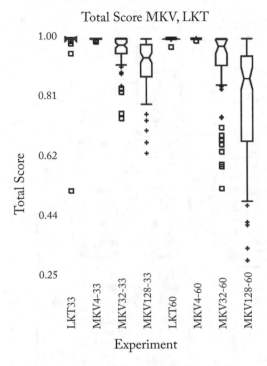

Figure 3.14: Inflected Box plots for the total score assessment for experiments using Markov chains and 3:3 and 6:0 look-up tables. Recalling that look-up tables can be thought of as Markov chains with two probability levels, the total score decreases significantly and substantially as the number of available probability levels are increased. The experiments based on three of the opponent's actions and three of your own also receive higher total scores than those based on six of an opponent's actions.

two additional epochs of evolution were performed and the Markov chains became even more cooperative in these later epochs.

As the number of probability levels available are increased the diversity of behavior exhibited by Markov chain-based agents increases. This is clear in both the play profiles shown in Figure 3.13 and in the total score statistics shown in Figure 3.14. Recall that total scoreto is the area under an evolving population's fitness as a function of the number of generations of evolution. This means the statistic will be lower for a population whose final level of cooperation is high but which took a long time to become cooperative.

Figure 3.14 shows a clear and significant decrease in the total score statistic as the number of available probability levels are increase. Remembering that look-up tables are, in essence, Markov chains with two probability levels, the mixing of these two representations is appropriate. This change in the evolutionary trajectory based only on the granularity of the probability of

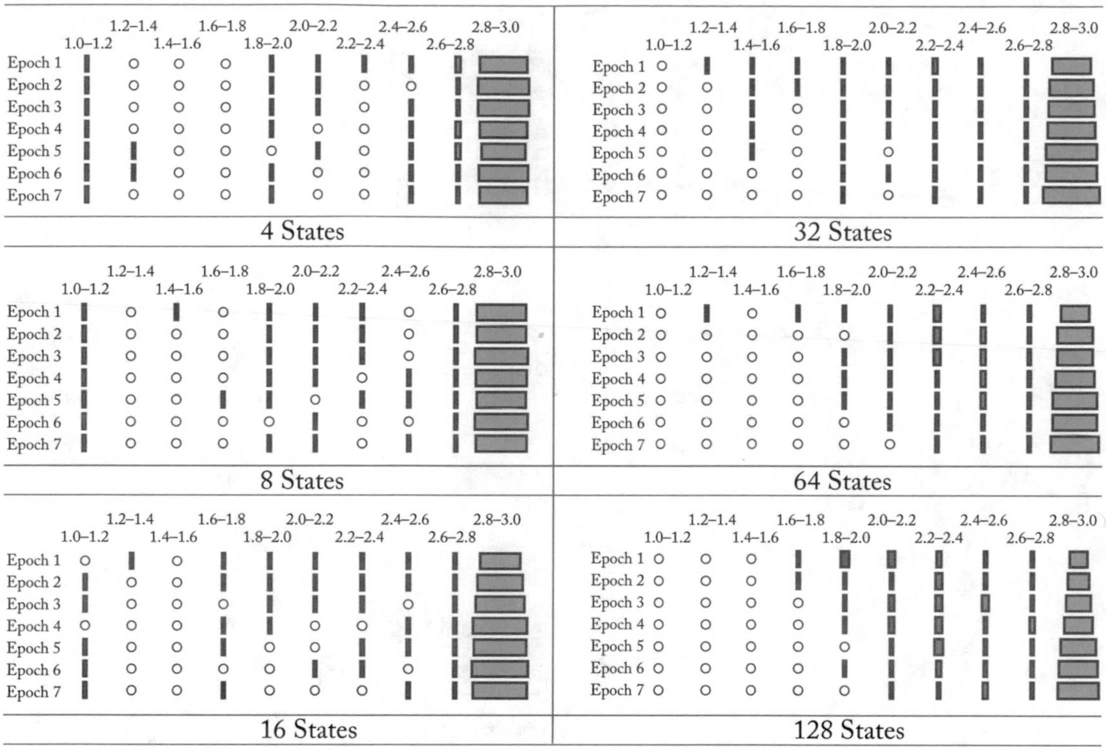

Figure 3.15: Play profiles for experiments using a finite state representation. The finite state representation becomes less cooperative and exhibits more diverse behavior as the number of states are increased.

cooperation was unexpected when the experiment was designed and suggests that other stochastic strategy representation may need more careful examination.

Figures 3.13 and 3.14 also show a clear effect of changing the input information resource. Markov chains with a 3:3 input structure are more cooperative and have a higher total fitness than those with the 6:0 input structure. Self-knowledge enables cooperation in Markov chain agents. This result is in exactly the opposite direction from look-up tables—although the resource variations are not directly comparable for the 2:2, 4:4, 4:0, and 8:0 input structure look-up tables.

The results for finite state machines, both the play profile data in Figure 3.15 and the total score information in Figure 3.16, show that 4 state agents are slightly less cooperative than 8 and 16 state agents. For agents with more than 16 states the level of cooperation exhibited in the play profiles and the total score statistics drop with the number of states.

The trajectory of the play profiles in Figure 3.15 shows that the different resource levels of FSMs are all becoming cooperative, but a different rates. This, in turn, shows the value of

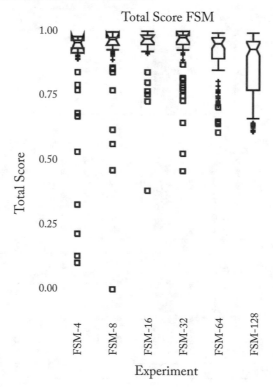

Figure 3.16: Inflected box plots for total score for the experiment run with finite state representations. The results echo experiment run with finite state representations. The results echo the results obtained with play profiles. The highest average score is for mutual cooperation, demonstrating that cooperation declines as the number of states increases.

the total score statistic. Figure 3.16 captures the slower approach to the cooperative state in a clear, clean fashion. The lower tail of outliers in Figure 3.16 also demonstrates that the agents with more available states have a narrower range of total scores, suggesting some type of stability resulting from increased resources.

Agents with 4–8 states do not change their level of cooperation much from one epoch to the next. Agents with 16 states become more cooperative over time and agents with more than 16 states both have levels or cooperation in the first epoch that are lower the more states are available but which rise over time more rapidly. The continued change in the 6th and 7th epochs suggest that running these experiments for an even larger number of epochs may see an additional increase in cooperation.

With other representations there is a meaningful way to change the amount of information directly available about past play. FSMs evolve and organize their own memory. This is one of

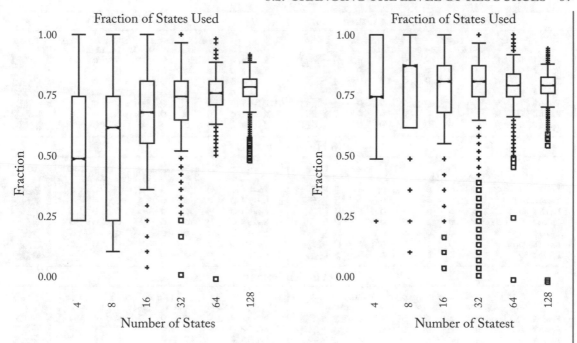

Figure 3.17: Inflected box plots for fraction of available states used by finite state agents. The left panel shows the results for evolved agents while the right shows the results for randomly generated agents. With the possible exception of 128 state agents, evolved agents use a higher fraction of their states than random ones. This demonstrates selective pressure to use states.

their strengths: where all three other representations must use a fixed memory depth and must spend space to encode a response for all possible patterns, finite state machines may reconfigure themselves to spend more resources on salient patterns in the opponent's behavior. The fraction-of-states used analysis was performed only for FSMs, in part to grant more perspective in place of that gained for other representations by examining input-information resources.

Figure 3.17 shows the distribution of the fraction of available states used in both evolved and randomly generated FSMs. Fraction of available, rather than absolute number of, states used was selected as an assessment to permit comparison between agents with different numbers of states. The results show that agents with fewer states available use significantly less states after evolution than when initially randomly generated; this difference decreases as the number of states available increases. Comparing evolved agents to one another, the fraction of states used increases in proportion to the number of states available, by a statistically significant margin. Given that the fraction-of-available-states statistic normalizes out the number of states available, this is an interesting result. It is also worth noting that the number of agents in the lower tail

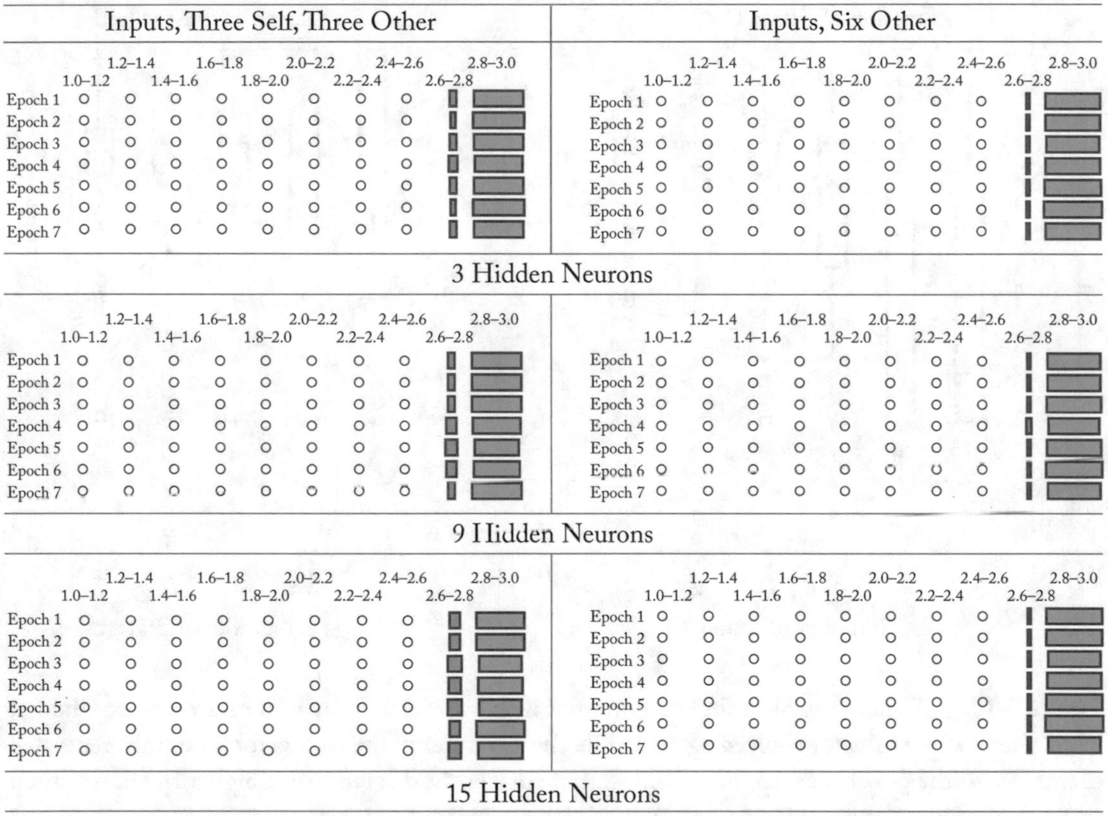

Figure 3.18: Play profiles for an experiment performed with ANNs. The left column shows results where the nets had three of their opponent's and three of their own actions as inputs. The right column shows results when the nets had the opponent's last six actions as inputs. As with the look-up tables, focusing on the opponent's actions increases the probability that cooperation emerges.

decreases under the effects of evolution. These data suggest that there is an arms race toward more complex strategies among the strategies with more available states.

The play profile in Figure 3.18 shows that, unlike Markov chains, ANNs are *more* cooperative when they have information only about their opponent. While there are statistically significant differences in the total fitness statistics for ANNs, those differences are small. The ANNs in this study are quite cooperative and, unlike the other three representations, do not get particularly more cooperative over time. The ANN results are the most stable with the possible exception of LUTs, but these are so cooperative that any variability is masked.

The box plots for the total score statistic for ANNs are shown in Figure 3.19. These demonstrate the stability of the neural representation when granted additional hidden neurons. There are significant differences present in the table, but they all arise from changing the input resources available to the agents. The 6:0 experiments which only possess knowledge of their opponent's moves are significantly more cooperative, as measured by total play, than the 3:3 agents.

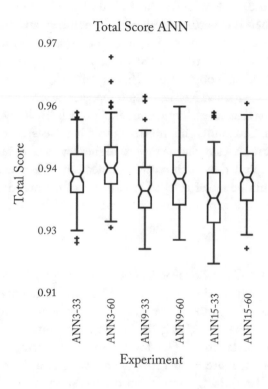

Figure 3.19: Inflected box plots for the total score assessment for experiments using ANNs. ANNs exhibited the least variation in total score of the changes in computational resources and available information.

Available Information vs. Representation   The finite state representation decides, through the structure of its state transition diagram, which of the opponent's actions it will remember or discard. It can also encode information about its past actions. The three non-finite-state representations were presented with either an equal mixture of the opponent's and the agent's own previous actions or an equal number of actions that were solely the opponent's. For the look-up tables and ANNs the experiments in which all the actions were the opponents exhibited a higher rate of emergence of cooperation. The Markov chains with a small number of probability levels mimic this behavior, but those with a large number of probability levels exhibited the opposite pattern of behavior.

The explanation for this probably lies in the granularity of the representations. An ANN or look-up tables encode a strategy with no randomness, a deterministic input-output map. The Markov chain representation uses random numbers while it is running its strategy. The more probability levels a Markov chain has available, the finer grained the representation. The fine-grained representation is more able to defect "just a little more." If the agent has access to its own actions, it can deduce when it is getting away with something, and so is tempted to continue looking for chances to defect when the opponent cooperates. If your only information is what the opponent is doing, then it is harder to tell when you are managing to collect temptation payoffs, yielding higher levels of cooperation. This explains the higher levels of cooperation by everything that is not a fine-grained Markov chain representation.

With Markov chains that have a large number of probability levels, the chance of defecting can change by a very small amount; this means that it is far easier to sneak in an occasional defection. This, in turn, causes a slow arms race to higher levels of defection as well as selective pressure to resist being exploited by such creeping defection. This means that the fine grained representation should exhibit a lower level of emergence of cooperation until defenses against creeping defection evolve. This explanation is consistent with the observed levels of emergence of cooperation.

Competitiveness Results   Collections of 10,000 pairs of agents from each of two resource levels were tested for competitive advantage for the FSM and MKV agents. The goal was to see which level of a given resource granted a competitive advantage. The results for the finite state agents were simple and conclusive—giving the agents more states granted a clear competitive advantage with a p-value better that 0.001 for a Z-test comparing the means of a binomial variable. This means that the more cooperative agents were at a disadvantage against the less cooperative agents, not too surprising a result, but there is another issue here, the *defensive value* of an agent.

Consider the three archetypes: AllD, AllC, and TFT. AllD is a mean agent; AllC and TFT are cooperative. They differ in that AllC is completely exploitable while TFT moves to defend itself quickly. Since the FSM agents with more states are slowly becoming cooperative while the agents with fewer states start cooperative, it seems quite likely that the agents with more states have more nuanced strategies and so have a higher defensive value. This means they are more likely to be able to exploit the agents with fewer states—a notion consistent with the outcome of the competitive analysis.

The results for the Markov chain agents were not simple. Figure 3.20 shows 95% confidence intervals for the probabilities that agents with more probability levels available would beat ones with fewer probability levels available in each of the seven epochs where agents were saved. The results for both 6:0 and 3:3 agents are shown. The probability level resource had a different impact in different epochs and behaved differently between the two Markov agent types tested. Early in evolution, agents with fewer probability levels played even or had a statistically significant competitive advantage. In the last epoch the agents with more probability levels always had a statistically significant advantage.

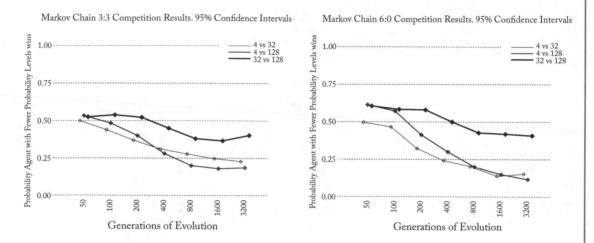

Figure 3.20: The two panels show the pairwise competition results for Markov chain agents. The left panel gives the results for 3:3 agents while the right gives results for 6:0 agents. While the plots show 95% confidence intervals, the use of 10,000 sample pairs of agents reduced the width of the intervals to less than a pixel in the above depictions.

It is intuitive that the Markov agents with fewer probability levels available can evolve faster; they are situated in a far smaller space of strategies. The space of agents with fewer probability levels are roughly a subspace of the gene space for agents with more probability levels. This means that, modulo the difficulty of evolutionary search the agents with more probability levels should gain an advantage in the end. On the other hand, the agents with fewer probability levels can probably *change* their strategies faster. This suggests that evolving agents with different probability levels *in competition with one another* would be an interesting experiment. The plots shown in Figure 3.20 contain examples of motion back to a lower level of competitive disadvantage for the 3:3 agents. This also suggests that none of the three agent types have yet completed their evolutionary arms race.

Nonlinear Projection   Figure 3.21 shows a nonlinear projection of the play profiles for FSM agents using all seven epochs and all numbers of states tested. The progression toward higher levels of cooperation visible in Figure 3.15 is also visible in the nonlinear projection. The higher the number of states, the less cooperative the agents are in the first epoch. The projection demonstrates that the FSMs with larger number of states also end up less cooperative. The 4 and 8 state machines and the later epochs for the 16 state agents are in the portion of the projection where the more cooperative play profiles are located.

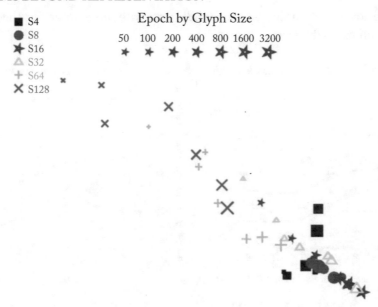

Figure 3.21: Non linear projection of the play profiles for finite state machines. Glyph sizes increase from the first to last epoch with different shapes and colors of glyphs denoting different numbers of available states. The experiments with more states move a greater distance in the projection than those with fewer states.

Most of the variation shown in Figure 3.21 in along a single axis running from less toward more cooperation. This suggests that the majority of change is the same sort of shift toward the sort of cooperative behavior exhibited by the eight state agents. The exception to this is the 4-state agents, which vary in an almost orthogonal direction. Looking back at at the play profiles shown in Figure 3.15, this is because of the large number of 4 state machines in the second bins with average scores in the range $2.6 \leq S < 2.8$. The nonlinear projection highlights the fact the 4 state machines are not part of the general trend.

3.2.3   CONCLUSIONS AND NEXT STEPS

This section extends the discovery of the impact of representation seen in [4, 21] to the behavior of evolved game playing agents. It is not necessary to change the representation to induce a significant change in agent behavior. Changing computational resources or information made available within a single representational type can also induce significant changes in agent behavior. While all four representations exhibited a response to changing available computational resources, Markov chains and finite state machines showed the largest differences. ANNs exhibited the smallest response to changing computational resources. These results are imprecise, however. The equivalence between additional states and additional neurons or probability levels is quite obscure. This study demonstrates a number of effects without being able to precisely compare them. Nevertheless, the first hypothesis, that changing computational resources induces changed in the behavior of evolved agents, is strongly confirmed.

The second hypothesis, that changing the information available to agents, is also confirmed. Having information solely about the opponent's past actions, rather than information divided between the opponent's and the agent's own actions, changes the level of cooperation, but in different directions for different representations. Most representations became more cooperative when presented only with information about their opponent, with MKVs. with a large number of probability levels being the exception. Experiments with LUTs increasing the amount of input information available made agents more cooperative.

In general, agents with more computational resources were less cooperative with the exception of the 6:0 neural nets which did not change significantly as more neurons were added. Since adding resources increases the strategy space represented, the lower levels of cooperation are almost certainly due to increase in the size of the search space. The cooperative agents strategies present in at the lower resource levels are still present at the higher resource levels and so available to be selected by evolution. The fact that they were not selected suggests they were lost in the larger search space.

In [21], the experiments were run for 250 generations while the current study ran agents for 3,200 generations. Representations, such as MKVs, which were assessed as relatively uncooperative in the earlier study became relatively cooperative in the last epoch of this study. Agent populations in which cooperation had emerged exceeded 50% of the total in epoch 3,200 in all the experiments. This suggests that increased resources do not prevent the emergence of cooperation, rather they delay it.

The Irrelevance of the Strategy Space   We note that both look-up tables and ANNs encode deterministic mappings from inputs to an output; in fact, they encode exactly the same set of potential phenotypes. Their play profiles are quite different. FSMs encode a space of strategies that contains the ANN/LKT space as a proper subset and Markov chains, being stochastic, encode a space of strategies substantially unrelated to the other representations. In spite of this, the four probability level Markov chains act almost exactly like look-up tables. In [8], several additional representations that encoded the strategy space of look-up tables were tested. Similarly in [4], two alternate encodings for the strategy space for FSMs were studied. The bottom line in this and the earlier studies is that encoding the same strategy space has nugatory value in predicting the level of cooperation, play profile, or behavior of evolving agents.

The representation and variation operators control the shape of the fitness landscape. The topology of the fitness landscape has far more impact on the behavior of evolved agents than the space of strategies encoded by a representation. As we increase resources within a representation, the space of strategies encoded by the representation expands and, with that expansion, a greater diversity of behavior emerges for most representations. The slower emergence of cooperation can be characterized as waiting to burn away enough diversity that the cooperative strategies can take over the population.

Revisiting Neural Nets   The most robust representation examined in this study were the ANNs. Changing the available input variables created a small, if statistically significant, effect in the total score statistic, as can be seen in Figure 3.19. There was only a modest effect from changing the number of neurons. This may be because the total range over which the number of neurons was changed was quite small and it may be because a single type of network topology was used. In addition, the original study on representation [21] uses two types of step-function neurons, one with a transition at argument zero and the other biased toward the cooperative output. Including different neuron types is another type of resource not examined in this study. All three changes, increasing the range for the number of neurons, studying other network topologies, and permitting variation in the type of neuron could be treated in future experiments.

Markov Competition   We report in this section interesting results for the evolution of the competitive ability of the Markov chain agents with different numbers of probability levels. It also speculates about the causes of these results, but well supported explanations will require additional work. In particular, the speed of adaption of agents with a different number of probability levels available should be checked. This can be done in the setting of *optimization Prisoner's Dilemma* in which agents are evolved to play against fixed collections of strategies. The evolutionary time required to find optimal average fitness against different sets of players can be used to characterize the adaptation rate of a representation in a given evolutionary setting. This, in turn, may yield an explanation for the reversal of advantage in which a small number of probability levels are superior early in evolution and inferior later in evolution.

Other Representations, Other Resources   We examine four representations, three if one considers that look-up tables are an extremely restrictive type of Markov Chain. The first representation study and its follow-up paper examined ten representations and subsequent papers have examined others. These included standard and linear genetic programming representations with and without internal state memory. Each of these representations has resources that can be varied. These resources include number of instructions in a generative FSM representation, number of nodes in a linear genetic programming representation, and the already mentioned variations in the neural nets.

3.3   ALGORITHM DETAILS MATTER TO POPULATION SIZE, ELITE FRACTION, AND MUTATION RATE

In many studies, the choice of agent representation is made without much concern and sensitivity to the choice of agent representation is ignored. Initial work on this issue appears in Chapter 2 which demonstrated that the choice of agent representation has a dominant effect on the probability agents will learn to cooperate. A theoretical treatment of why some representations are more likely to yield cooperative behavior appears in [34]. This line of research was extended by examining the impact of simply varying the resources allocated to agents within a representation in Section 3.2. Changing the number of states in a finite state representation, the number of probability levels available to a Markov chain, the number of neurons in an ANN, and the time depth of a look-up table were all found to change the behavior of agents substantially.

Taken together, the studies in this book demonstrate that our understanding of the interaction of the way agents are encoded and the process of training those agents with evolution is so incomplete as to make research that uses evolved game-playing agents unreliable. The goal of using IPD-playing agents as models for cooperation and conflict both for humans and animal communities requires that we complete the characterization of evolutionary training of agents for playing the IPD.

Previous sections that examined representation and agent resources were painted in relatively broad stokes. While this section retains variation of a resource as part of its experimental design, the number of states available to an agent, its primary purpose is to narrow the brush strokes to the algorithm parameters elite size and population size in a manner that can be compared with other studies in this series. As we will see, even changing the simple, basic algorithm parameters changes the type of agents that arise under evolution. This study may complete the identification of factors *other than the problem modeled* that affect the outcome of experiments using evolved IPD agents. At present, published studies are likely to fail to state important experimental design parameters because those factors were not yet clearly identified. This claim of completeness also does not apply when additional factors like imposing geography or additional game rules are included.

We will perform this examination using the direct finite state agent representation because it is the most studied representation. We retain the agent-training algorithm from earlier studies but modify parameters held constant in other studies. The standard algorithm operates on a fixed size population with 30 or 36 agents. One-third of the agents are replaced in each population updating. This is called having an *elite fraction* of two-thirds—the elite are those agents not replaced. These factors were held constant while others, such as agent representation or resources made available to an agent, were changed. In this section we permit population size and elite fraction to vary, examining nine new pairs of values for these factors. This study uses a finite state representation for agents, a representation that has exhibited diverse behavior in the past [7]. The experiments look at agents with a small, standard, and large number of states. Each of these three factors is examined to see what impact it has on agent play behaviors.

3.3.1   DESIGN OF EXPERIMENTS

The finite state machines used in section use the *Mealy* architecture with responses encoded on the transitions (the Moore architecture encodes responses on the states). An example of a Mealy machine appears in Figure 3.22. Transitions in the finite state machine used in this study are driven by the opponent's last action and so an initial action is also required. Machines are stored as a string of states with a state incorporating both transitions and responses associated with the state. The initial action is stored with the first state. Crossover operates on the string of states; this study uses two point crossover. The mutation operator operates by changing the initial state or action 5% of the time (each), the destination of a state transition 40% of the time, and a response to an opponent's action 40% of the time. This mutation operator is retained from previous studies for consistency.

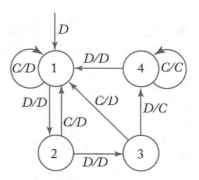

Figure 3.22: An example 4-state finite state machine for playing IPD. Transitions are labeled as **input/output** and the sourceless arrow in the upper right is labeled with the machine's initial action.

We examine 27 sets of algorithm parameters. These experiments form a full factorial over the values population size {72, 108, 144}, elite fraction {1/3, 1/2, 2/3}, and numbers of states {4, 16, 128}. For each set of parameters, 100 sets of evolutionary runs are performed for 3,200 generations. In each generation the members of the population play a round-robin tournament and the average fitness against all opponents is computed. This fitness is used for a fitness-proportional selection of pairs of parents from the elite to replace the non-elite portion of the population. Pairs of parents are copied, the copies undergo two-point crossover of their string of states, and then the copies are mutated to generate the children which are placed into the population.

This design of experiments, together with the analysis techniques described subsequently, will permit us to assess the way that the three factors—population size, elite fraction, and number of states—interact to affect the types of agents that arise.

Analysis Techniques   We retain two analysis techniques from Section 3.2, play profiles and the total score statistic. The ability of agents trained using different algorithm to compete in one-on-one contests is also assessed. Sampled over a large number of pairs of agents, the Bernoulli variable "who won" can be used to estimate the probability of victory of one agent type over another.

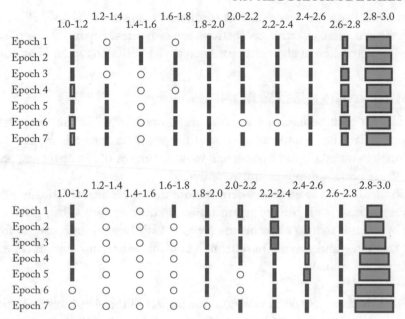

Figure 3.23: Two examples of play profile for 3,200 generations of evolution, partitioning the possible population average scores into ten bins. Two examples were used to display all possible colors. Colors are explained in the text.

Competitive Ability The competitive ability of agents with different numbers of states, against one another, was assessed in the following manner. For each pair of numbers of states, agents were sampled randomly from the elite population saved in epoch 7 (at 3,200 generations) for each experiment. These agents played 150 rounds of IPD. If one agents total score was 5.0 points higher that agent was declared the winner, otherwise the agents were deemed to have (approximately) tied. This means that entirely cooperative agents were washed out of the sampling process.

Sampling of pairs of agents was continued until 400 decisive results were achieved. A 99% confidence interval for the probability of victory by the agent with a larger number of states, given that the agents did not tie, was computed using the normal approximation to the binomial. If v is the number of agents with more states that won and n is the number of samples this confidence interval is given by:

$$\frac{v}{n} \pm 2.12 \times \sqrt{\frac{v/n \cdot (1 - v/n)}{n}} . \tag{3.4}$$

Two sets of agent trained in different ways are presumed to have significantly different competitive ability if the confidence interval does not include $p = 0.5$.

Since there are contests where agents compete to play the IPD [39, 79], the algorithm parameters that yield more effective competitors are of interest beyond the application in this study of showing that different algorithm parameters yield different agent behaviors.

3.3.2   RESULTS AND DISCUSSION

The play profiles for all experiments are shown in Figure 3.24 while the total fitness statistics appear in Figure 3.25. The assessment of competitive ability is shown in Figure 3.26. None of the three parameters varied display a consistent trend for any of the assessments, demonstrating that nonlinearity is present throughout the system.

In addition, all three assessments demonstrate that there are statistically significant differences between at least some pairs of agents trained in different ways. This gives an affirmative answer to the hypothesis that experiments using evolved IPD agents must control for population size and elite fraction—and also the number of a available states, although this factor has been investigated in earlier studies.

Impact of Elite Fraction   So far as we know, the impact of the elite fraction in the context of training IPD-playing agents has not been previously studied. Since replacing a lower fraction of agents reduces the rate of change of agent behaviors the expectation is that higher elite fractions would yield more cooperative agents. Figure 3.24 supports this hypothesis for the 4 and 16 state agents, but not for the 128 state agents. For the smallest population size, 72, the populations with the lowest elite fraction become *more* cooperative.

This study confirms results in earlier studies with population size 36 that agent populations of 128-state FSMs become cooperative more slowly than agents with fewer states. If we view low elite fractions as increasing the rate of evolution—plausible as a low elite size increases the rate of introduction of new types—then these results are consistent with the following explanation. The lower elite size increases the rate of evolution, arriving at a more cooperative state sooner.

The phenomenon of higher elite fractions encouraging cooperation are most pronounced for the populations of 4-state agents. Its relative subtlety for the 16-state agents suggests that the two effects, encouraging cooperation by having a more predictable agent population and speeding the emergence of cooperation as observed clearly in the 128-state agents, are in a state of near balance for the sixteen state agents.

The total fitness statistics also show a more pronounced increase in cooperation for the 4-state agents than for the 16-state agents; the 16-state agents have better total fitness statistics overall, perhaps masking the improvement. To total fitness statistics for the 128-state machines show lower values, but also less of a trend associated with elite size, supporting the results from the play profiles.

Examining Figure 3.26, we see that, while agents with less states are uniformly outcompeted, the margin of victory for the machines with more states is smaller when the agents with fewer states have a lower elite fraction that the agents with more states. If evolution tends

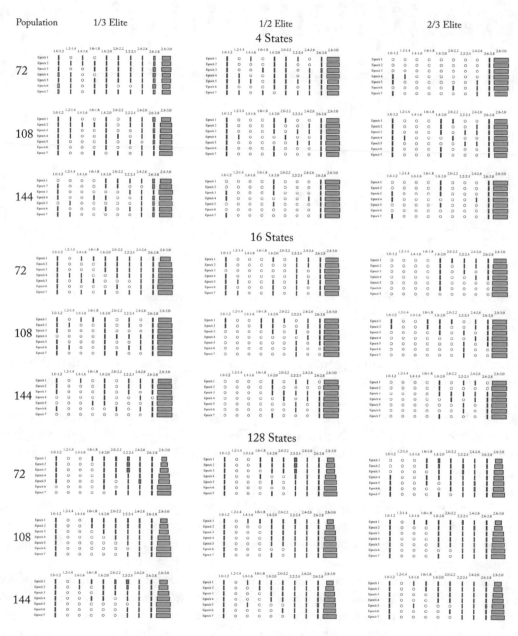

Figure 3.24: Play profiles for all 27 experiments performed.

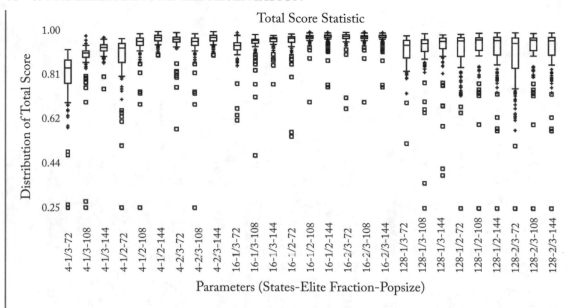

Figure 3.25: The distribution, over 100 runs for each set of parameters, of the total fitness statistic for all 27 experiments performed.

to grant more competitive ability, something that was found to be true in [25], then this observation also supports the hypothesis that smaller elite fractions permit more evolution.

Impact of Population Size   For the 4- and 16-state agents, larger populations clearly encourage cooperation. This is visible in both the play profiles shown in Figure 3.24 and the total fitness statistic in Figure 3.25. These figures also provide support for this trend in the elite-fraction 1/3 populations of 128-state agents, but they exhibit a weaker increase in cooperation. The two higher elite fraction show no such trend for the 128-state agents.

The major impact of having 128 states available to the agents is to astronomically increase the number of available strategies. In particular, this increases the potential for *hand-shaking* strategies [37]. These strategies engage is brief, low-fitness play as a means of performing kin-recognition.

These results are in contrast to a study that used a large, geographically structured population [35]. In this study, large populations rapidly evolved to a highly cooperative state. This study differed both in using a geographically structured scheme for agent breeding and a sampled, as opposed to round-robin, fitness evaluation scheme. This suggests that both factors, geographic structure and sampled fitness evaluation, merit additional investigation.

Population size had a modest impact on competitive ability. When there is an observable effect, the winning agent population does better when the loosing agent population is larger.

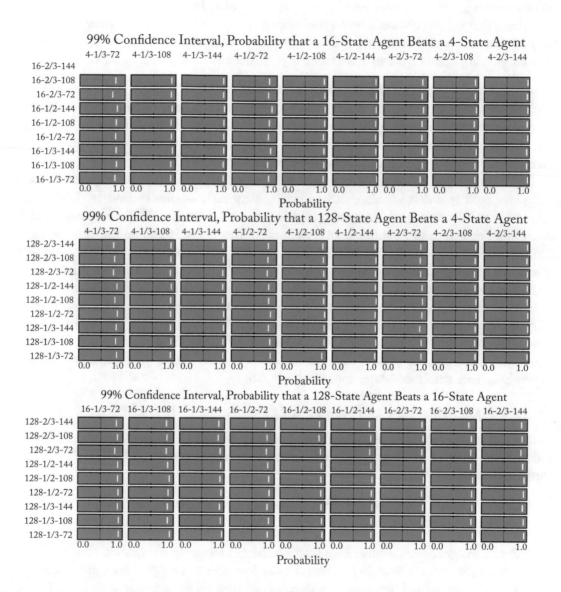

Figure 3.26: 95% confidence intervals on the probability that an agent with one number of states beats an agent with a different number of states in 150 rounds of IPD. All nine experiments with each sort of agent are compared. Labels are of the form states-elite fraction-population size. The probability is denoted by the position of the white bar with the confidence interval given by the width of the shaded region about that bar.

Since larger populations evolve more slowly when learning IPD, this is more support for the hypothesis that more evolution is better for competitive ability. The jump-start one might expect from having a more diverse initial population is apparently absent—but this is not too surprising as game-playing agents typically have an early bottleneck in which most agent types die out.

Impact of Available States   Of the three factors examined in this study, the number of states available to agents is the one for which the most work has been done recently after noticing the anomalous results of changing the number of states in [7]. The strongest result in this study is that changing the number of states modifies the impact of both the other factors varied. Having a large number of states, as noted above, reverses the impact of the elite fraction.

All three numbers of states examined exhibited different behaviors. In past studies, 16 states was used as a default value, dating back to very early studies such as [85]. This study detected that 16 states is within a balance point where the impact of elite size and available states more or less cancel out one another's impact. This is ironic and re-enforces the notion that there is a need to more fully characterize the behavior of systems of evolving game-playing agents before using these tools with confidence as models of cooperation and conflict.

The competitiveness results demonstrated that agents with more states have a clear competitive advantage. In 81 comparisons between 4-state and 128-state agents, the lowest probability of victory by 128-state was 0.7725 ± 0.0444, for 2/3 elite and population size of 72 for the 128-state agents and 1/3 elite with a population size of 72 for the 4-state agents. Figure 3.26 shows that there were no pairs of experiments where the confidence interval on probability of victory by agents with more states contained $p = 0.5$.

3.3.3   CONCLUSIONS AND NEXT STEPS

In this section we clearly demonstrated an interaction between elite fraction, population size, and number of states made available. Two impacts of elite-fraction were detected that pull in opposite directions.

- When agents are small enough (few states) to evolve efficiently, higher elite fractions encourage cooperation by making the rate at which the population changes slower.

- For large agent sizes, smaller elite fractions permit more rapid evolution of agents. These larger agents were found, in previous studies, to evolve to a cooperative state more slowly. This, in turn, caused *lower* elite fractions to encourage cooperation.

We also showed that for agents with few states, larger population sizes encourage cooperation; this effect was not visible for agents with larger numbers of states. The study also confirmed the results of earlier studies on the relatively slow emergence of cooperation in agents with more states.

One clear outcome of the study in this section is that elite fraction, or its equivalent in other sorts of training algorithms, as well as population size, need to be added to the list of parameters that influence the outcome of training game-playing agents with an evolutionary algorithm.

A Protocol for Reporting Evolution of Game-Playing Agents   The material presented here is part of a fairly large collection of research that examine uncontrolled sources of variation in experiments on the evolution of game-playing agents. An early priority for future research is to compile a careful list of factors that matter and draft a proposed set of reporting criteria for this sort of agent based study. Such a proposal must be a draft and would require input from a broad variety of researchers to become acceptable for adoption by the research community.

Evolutionary Time and Lineage   Another interesting effect found here is the impact of both number of states and time spent evolving on an agent's competitive ability. The quantity "time spent evolving" needs to be examined with greater care. The number of generations a population is permitted to evolve is a course measure of the amount of evolution that has taken place. A better measure might be *average lineage length*.

Definition 3.1   A **lineage** is a chain of descent with nodes that are members of an evolving population and links that join a parent and child. The **lineage length** of an individual is the longest lineage for which that individual is the endpoint.

Notice that a lineage may be part of a larger lineage—a population member that appears in the second generation is the end of a lineage of length one, even if it goes on to have its own children with longer lineages that descend through it.

We propose lineage length as an assessment for future work because it seems a more accurate means of measuring the amount of evolution that takes place. If, for example, an elite population remained the same for many generations, producing inferior children, then it would have an average lineage length shorter than its time of evolution as measured in generations. Comparing lineage length with level of cooperation and competitive ability may yield cleaner results than measuring evolutionary time in generations.

Note that the lineage length of an individual is simply one plus the larger of its parent's lineage lengths; this means that the entire pattern of descent need not be saved making this proposed assessment relatively inexpensive in both space and computing power.

Deep Time Studies   In [37] is was found that new strategies, not previously seen, arose in the gap between generations 32,768 and 65,536 (these were the last to sampling epochs) in populations of 20 state finite state agents playing the IPD. Given the apparently slower rate of adaption in agents with 128 states, this suggests that even deeper time studies may yield new, complex strategies in agents with large numbers of states.

Other Representations and Algorithms   In this section we demonstrate that directly encoded finite state agents exhibit a sensitivity to the choice of population size and elite fraction. As noted in the Introduction, there are many other representations that can be used to evolve agents to play

IPD. There are also may be games beside IPD that are potentially interesting to train agents for. This suggests a vast cone of future research projects checking other representations and games.

We have also stuck with a particular agent training algorithm used in several other studies in the name of controlling sources of variation. The use of a round-robin tournament to assess fitness, the use of an elite for reproduction, and the absence of a hall of fame or other source of long-term genetic memory other than direct inheritance are all features that still have not been investigated.

Geography   Two earlier studies [25, 35] examined the evolution of finite state agents for the IPD using very different geographic structures. In both studies the diversity of agent types were found to have been enhanced. In the latter study, it was found that the particular type of geographic structure with local breeding but global fitness evaluation, encourages a robust form of cooperation. The agents not only cooperated with one another, but exhibited an enhanced resistance to invasion. These studies suggest that it may be worth structuring an investigation around the following questions.

- What is the impact of having and varying a geographic population structure?

- What is the impact of a round robin, geographically local, or geographically random choice of opponents for fitness evaluation?

- Do disturbances—disruptions that remove players from parts of the geography—have an impact on the types of agents that evolve?

This latter question dovetails with ecological investigations on the importance of disruptions to ecology and bio-diversity [81, 105].

3.4   INCLUDING TAGS AND GEOGRAPHY

This section serves to introduce ideas about noise that will be needed in Chapter 5. It also dips into two subjects deserving of additional study: using geography to structure agent interaction and giving agents a way of telling one another apart (tags).

It is generally believed that evolution is an undirected process yielding only solutions appropriate to the current environment rather than general adaptive features. Despite this general consensus, some researchers have questioned this belief. One example of this questioning comes from Dawkins [52] when he writes

"If ... predators from one era could meet prey from another era, the later, more 'modern' animals, whether predators or prey, would run rings around the earlier ones. This is not an experiment that can ever be done ..."

Although using biological organisms for such a study is not practical, the use of an evolutionary algorithm to conduct such a test is possible and has been done. These studies have found that populations from a more modern era can consistently out-compete populations from a more archaic era and this effect holds even when the populations come from different worlds [9, 25]. We call this phenomenon *non-localized adaptation*.

The goal of this section is to take a closer look at non-localized adaptation and observe the effect of agent recognition on this phenomenon through the use of a model where non-localized adaptation has been previously demonstrated. In the original model, a population of 100,000 agents played their neighbors for 150 rounds of *noisy Iterated Prisoner's Dilemma* (nIPD). The agents were placed on a 100×100 toroidal grid. The grid was run for 10,000 generations with a mutational form of evolution running and the grid was saved at generation 100 and 10,000. The grid was loaded and run 30 times yielding 30 archaic (generation 1,000) and 30 modern (generation 10,000) populations.

Each combination of modern vs. archaic was allowed to compete for 100 generations with no new strategies generated. At the end of the 100 generations, the number of modern and archaic agents were counted. In seven different variations on this experiment it was found that the modern agents were able to significantly out-compete the archaic agents in all cases [25]. These results demonstrated that the agents were evolving a general ability to play nIPD in addition to specific adaptation to their own populations.

Given a model of non-localized adaptation it is natural to ask questions of that model to obtain a deeper understanding of the phenomena. In this study we make measurements at more frequent intervals to get a view of the rate of non-localized adaptation throughout the initial 10,000 generations of a population. We alter the rate of mutation in the population to identify the effect of mutation rate on non-localized adaptation. We also study a generalization of the model by adding a rudimentary agent recognition system to examine the effect of increased complexity on non-localized adaptation.

3.4.1   GRID-BASED NIPD AND AGENT SPECIFICATIONS

The agent task for this study is nIPD. In IPD the same two players play against each other repeatedly. These multiple iterations allow for complex play to develop. With this more complex play, interesting strategies emerge. One example of such a strategy is called tit-for-tat. In this strategy, the agent's first move is C and every subsequent move is the same as its opponent's last move. In IPD (without noise) this strategy emerges as a particularly good solution. To eliminate this preference we introduce a small chance ($\alpha = 0.01$) that an agents move will be replaced by the opposite move. With the addition of noise, tit-for-tat is transformed from an excellent strategy to one that is no better than totally random play when playing against a copy of itself for many iterations.

Definition 3.2   An agent consists of an identifying tag, an array for handling opponent tag information, and a finite state machine (FSM) adapted to play nIPD in Figure 3.27. The tag value is a number that serves as a label that is exchanged at the beginning of play, there is no explicit

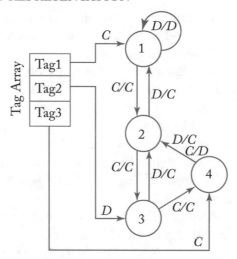

Figure 3.27: Diagram of a randomly selected agent with tag recognition modifications. The tag type of this agent is not specified. The tag array gives potentially distinct initial states for each possible tag that may appear on an opponent.

attempt to give meaning to the tags (i.e., associating them with a particular FSM arrangement) rather the tags are allowed to take whatever meaning that may arise through evolution in the population.

Agent play begins by looking up the appropriate initial action and state in the tag array for the opponent's tag number. The agents then make their initial move and set their current state. Each state has two outbound transitions, one for each of the opponent's possible moves. Each transition gives the action the agent should take for the next round of play and points to the next internal state.

Example 3.3   Two agents with the FSM pictured in Figure 3.27 but one had a tag value of 1 (agent a) and the other a tag value of 2 (agent b) play would go like this: The agents would read each others tags, set their initial state and make their first move. Play would be as follows.

Agent A			Agent B		
Score	State	Move	Score	State	Move
0	3	D	0	1	C
5	4	C	0	1	D
5	2	C	5	2	C
8	3	C	8	3	C
11	4	C	11	4	C

Agent growth can happen through a number of mutations. A growth mutation consists of randomly selecting a state in the FSM, copying it, and randomly changing some of the inbound transitions of the original state to the new state. A contraction mutation consists of selecting a state at random and deleting it. Inbound connections to the deleted state are randomly assigned to the remaining states in the FSM. A basic mutation consists of randomly selecting a transition and then changing either the state that it points to or the agent's action for that transition. A tag mutation consists of changing the agent's tag to a new randomly selected value. A tag array mutation involves selecting one of the tag array positions and changing either the initial action or changing initial state for that tag to a state selected uniformly at random from those available. The rates of mutation used for individual experiments are given in the descriptions of the experiments later in this section.

The world consists of a 100×100 toroid shaped grid. The Toroidal topology is chosen to eliminate edge effects that can arise using a simple grid topology. The grid is initially populated by randomly generating 16 agents and filling the grid with randomly selected copes of these agents. Agents play their neighbors to their North, South, East, and West. A generation consists of all grid positions playing their neighbors followed by a selection step. The score for an agent in a particular position in the grid consists of the average score obtained from playing all neighbors. Rather than playing all 1,200,000 pairwise games of IPD needed for a generation, the score for a pairing is computed using a dynamically generated score matrix where position $x(i, j)$ is the average score of 5 sessions of nIPD where each session consists of 60 rounds of Prisoner's Dilemma. We choose 60 rounds for the session length as it seems to be long enough to allow for complex play to develop and for non-local adaptation to occur. The average score from 5 sessions is used to provide a more representative score for the pairing as the score is use throughout the grid wherever that pairing occurs. A new matrix is generated for each generation and the matrix is indexed by the agents present in that given generation. After a generation of play each position in the grid is updated by copying the highest scoring agent for the neighborhood into the grid position or leaving the agent currently occupying the space in place if it outscored all of its neighbors. During the update there is a small chance (1 in 100,000 for most of the experiments here) that a mutation will occur. For each condition 30 populations are generated, each population is allowed to evolve for 10,000 generations, the grid is saved every 100 generations.

A competition grid is created by loading one half of a grid with the corresponding portions of the saved grid from the archaic generation and the other half with the corresponding portions of the saved grid from the modern generation. Mutation is turned off and the grid is run for 100 generations. The number of modern and archaic agents is counted; if the number of modern agents is greater than the number of archaic agents it is considered a win for the modern population. This competition is performed for every pairing of the saved modern and archaic populations, resulting in 900 trials. From this we can calculate win percentage of the modern generation. A 99% confidence interval is then calculated for the win percentage using a normal approximation to the binomial distribution.

As one of the major objectives of this study is to compare and contrast non-localized adaptation with and without tags, there was a need to generate populations without tags. For this study, populations without tags were accomplished by setting the maximum label value to 1. With this setting the agents functionally identical to the tagless agents in the previous experiment. This similarity is further confirmed by the similar results achieved for the tagless populations between the results for this study and the previous study.

3.4.2   EXPERIMENTAL DESIGN

We performed three different experiments in this section. The first took a closer look at non-local adaptation by looking at more pairs of distinct generations than the initial study. We also performed a run with tags introduced at low mutation levels to get an initial feel for the effect of tags on non-localized adaptation. The next experiment varied the rate of mutation of both agent tags and their tag arrays to determine what effect this had on non-localized adaptation. The final experiment varied the total mutation rate for populations with no tags to identify what effect the overall mutation rate had on non-localized adaptation.

A Higher Resolution Look at Non-localized Adaptation   In order to obtain a closer look at non-localized adaptation we repeated the "null environment" experiment from [25]. In this run the only parameter that was different was the frequency with which the grid was saved. As mentioned previously, for our data every 100th generation was saved. Due to the computational time required to perform all 900 competitions required for a comparison of generations, we selected every 500 generations for comparison. This yielded a measurement of non-localized adaptation for every 500th generation. For the agents with tags the relative mutation levels were 50 basic : 25 growth : 25 contraction.

In addition to the run without tags, we also performed a run with tags. For this run we selected an arbitrarily small tag mutation rate. For this run the relative mutation rates were 40 basic : 25 growth : 25 contraction : 5 tag : 5 tag array. The maximum tag value was 20, meaning that agents could have a tag value from 1–20. Tag values and tag arrays were generated randomly when the agents were created. Again the populations were allowed to evolve for 10,000 generations and every 500th generation was placed in competition against generation 10,000. For both data sets 99% confidence intervals were calculated using the previously described methods.

A Survey of Different Tag and Tag Array Mutation Levels   A better picture of the effect of tag mutation levels on non-localized adaptation was generated by varying tag mutation levels and tag array mutation levels over three levels. With this survey of tag related mutation rates, our goal was to identify if different combinations of these rates had different effects on non-localized adaptation occurring in the population.

To accomplish this comparison we performed nine data runs that differed only in the tag and tag array mutation levels. The three different levels were 10 (low), 25 (medium), and 50 (high). These values were chosen as they represented a relatively broad range of frequencies from less frequency than the FSM mutations (growth, basic and contraction) to higher than the FSM mutations. As before every 500th generation was placed in competition against generation 10,000, the percentage of modern "wins" was recorded, and the 99% confidence interval was calculated.

In order to get a better picture of how tags were being used by the populations, we calculated the number of agents displaying each particular tag for every 100th generation. With this data we can identify if there is a large diversity of tags in a given grid or if most of the agents tend to have the same tag.

A Look at Overall Mutation Rate   As the relative frequency of tag mutations increase, the result is a decrease in the frequency of FSM mutation decreases. This occurs in our experiments because we change the relative frequencies of the mutation types without changing the overall mutation rate. For example, if the relative rate of tag mutations equals the relative rate of FSM mutations, the result is that the rate of FSM mutations equals half that of the overall mutation rate. We performed a study to see what the effect of changing the overall mutation rate was. For this experiment we performed two additional runs. These runs were identical to the "no tags" data run with the exception of the overall mutation rate. The low mutation rate was set at 0.1× the rate of the "no tags" run, this meant that there was roughly 1 mutation event every 10 generations. The high mutation rate was set at 10x the rate of the "no tags" run. This meant that there were roughly 10 mutations per generation. Again, every 500th generation was placed in competition with generation 10,000, the percent modern generation wins were calculated, and the 99% confidence interval was generated.

3.4.3   RESULTS

The key test in this study was the competition between the archaic and modern generations. This competition was run for a number of populations with and without tags with varying mutation rates. Figure 3.28 is a graph that demonstrates the frequency with which the modern generation dominated the archaic generation for the initial tags/no-tags data runs. For all of these graphs, the x-axis represents the archaic generation that was placed in competition with generation

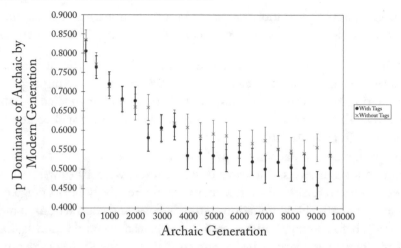

Figure 3.28: Graph showing the proportion of 900 trials where the modern generation (10,000) dominated the archaic generation (x-axis) for populations with tags (red x) and without tags (black dot). Error bars indicate 99% confidence interval.

10,000. The y-axis represents the proportion of the 900 trials where the modern generation dominated the archaic generation. The error bars represent the 99% confidence interval for the values. The table in Figure 3.29 holds the results for the 3 × 3 survey of tag mutation levels. The naming convention for the mutation levels is tag level + tag array level. For instance, the name medhig represents the population with the tag mutation level of 25 and the tag array mutation level of 50. Each number is prefaced by a symbol.

The key to the symbols used in the table of Figure 3.29 are as follows. The (⋆) values are significantly higher than the corresponding value from the experiment without tags based on the 99% CI, (+) values are higher than the corresponding value from the experiment without tags but is not significant based on the 99% CI, the (−) values are lower than the corresponding value from the experiment with no-tags but the value is not significant based on the 99% CI, the (#) values are lower than the corresponding value from the experiment without tags and is significant based on the 99% CI. Figure 3.30 is a graph of the results from the data runs where the overall mutation rate was changed. This graph follows the conventions of the first graph with the x-axis representing the archaic generation and the y-axis representing the proportion of the competitions that the modern generation dominated the archaic generation.

In the competitions most of the comparisons are between archaic and modern generations that come from different populations. This means that these populations have different evolutionary history that cannot have evolved to play against each other. Given this different evolutionary history, if no non-localized adaptation were to occur we would expect the modern generations to have no advantage over the archaic generations. For this study, this absence of

Archaic Gen.	notag	lowhig	lowmed	lowlow	medhig	medmed	medlow	highig	higmed	higlow
100	0.83	(#) 0.72	(+) 0.84	(-) 0.81	(#) 0.72	(#) 0.74	(+) 0.85	(+) 0.85	(#) 0.79	(-) 0.80
500	0.77	(#) 0.68	(*) 0.82	(+) 0.78	(#) 0.66	(#) 0.66	(-) 0.74	(-) 0.76	(#) 0.67	(-) 0.75
1000	0.71	(#) 0.59	(*) 0.76	(+) 0.73	(#) 0.59	(#) 0.61	(+) 0.72	(*) 0.78	(#) 0.65	(-) 0.70
1500	0.68	(#) 0.58	(*) 0.75	(+) 0.68	(#) 0.56	(#) 0.61	(+) 0.69	(-) 0.67	(#) 0.59	(+) 0.70
2000	0.66	(#) 0.62	(*) 0.71	(+) 0.68	(#) 0.53	(#) 0.62	(+) 0.67	(+) 0.68	(#) 0.60	(+) 0.69
2500	0.66	(#) 0.59	(+) 0.69	(*) 0.70	(#) 0.57	(#) 0.56	(+) 0.67	(-) 0.63	(#) 0.60	(+) 0.67
3000	0.6	(#) 0.56	(*) 0.67	(+) 0.64	(#) 0.55	(#) 0.56	(+) 0.62	(+) 0.62	(-) 0.60	(+) 0.63
3500	0.62	(#) 0.56	(*) 0.66	(+) 0.62	(#) 0.57	(#) 0.55	(+) 0.63	(-) 0.61	(-) 0.60	(+) 0.65
4000	0.61	(-) 0.59	(+) 0.63	(+) 0.61	(#) 0.53	(#) 0.52	(+) 0.61	(-) 0.57	(-) 0.59	(+) 0.62
4500	0.59	(-) 0.56	(+) 0.61	(*) 0.64	(#) 0.53	(-) 0.56	(+) 0.59	(+) 0.60	(-) 0.58	(+) 0.61
5000	0.59	(#) 0.54	(+) 0.62	(-) 0.57	(#) 0.53	(-) 0.59	(*) 0.63	(+) 0.60	(-) 0.59	(-) 0.58
5500	0.59	(-) 0.55	(+) 0.59	(+) 0.59	(#) 0.52	(-) 0.58	(*) 0.63	(+) 0.59	(-) 0.58	(-) 0.58
6000	0.56	(-) 0.55	(*) 0.62	(*) 0.61	(#) 0.50	(-) 0.55	(+) 0.59	(-) 0.55	(+) 0.57	(-) 0.56
6500	0.57	(#) 0.53	(*) 0.61	(+) 0.59	(#) 0.52	(#) 0.52	(+) 0.57	(-) 0.54	(-) 0.56	(-) 0.55
7000	0.57	(#) 0.54	(+) 0.60	(-) 0.55	(#) 0.53	(#) 0.50	(#) 0.52	(+) 0.57	(+) 0.53	(-) 0.55
7500	0.55	(#) 0.50	(*) 0.60	(+) 0.58	(#) 0.48	(#) 0.50	(#) 0.51	(-) 0.55	(-) 0.54	(#) 0.50
8000	0.55	(-) 0.53	(*) 0.59	(+) 0.57	(#) 0.49	(#) 0.47	(+) 0.55	(-) 0.52	(#) 0.50	(-) 0.53
8500	0.54	(-) 0.52	(*) 0.62	(*) 0.58	(#) 0.45	(#) 0.47	(+) 0.56	(-) 0.51	(#) 0.48	(-) 0.52
9000	0.56	(#) 0.51	(*) 0.61	(-) 0.54	(#) 0.50	(#) 0.51	(-) 0.53	(-) 0.53	(#) 0.48	(-) 0.53
9500	0.54	(+) 0.54	(-) 0.53	(-) 0.51	(-) 0.52	(#) 0.48	(-) 0.53	(-) 0.50	(#) 0.48	(+) 0.54

Figure 3.29: Results from 3 × 3 mutation level study.

non-localized adaptation would be seen as the 99% CI including the value 0.5. The initial study looked only at generation 100 vs. generation 10,000 and saw non-localized adaptation as defined by this test. Our goal is to take a closer look at these findings.

Non-localized Adaptation Over Time   The previous study only tested generation 100 against generation 10,000. This test found that the modern generation dominated the archaic generation a significant proportion of the time. Although these results are significant, one still wonders if non-localized adaptation can be seen at later points in evolution or if non-localized adaptation is something restricted to the relatively "young" evolving population.

In the initial data runs we see that non-localized adaptation is not restricted to the first few hundred generations. In the run that is a replication of the previous study, the 99% confidence interval never overlaps with 0.5. This implies that non-localized adaptation is occurring throughout the entire data run. Most impressively, the competition between generation 9,500 and generation 10,000 still proves to be significantly different from 0.5 (0.5356 ± 0.0353). This

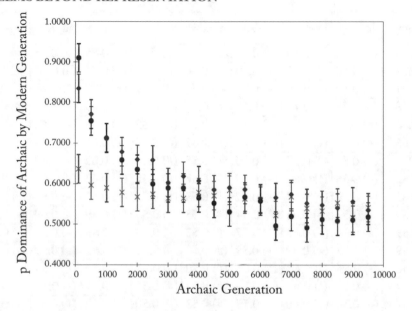

Figure 3.30: Graph showing the proportion of 900 trials where the modern generation (10,000) dominated the archaic generation (x-axis) for three different tagless populations. Diamond = populations at normal overall mutation rate, X = 0.1X, and circle = 10X.

implics that even at generation 9,500 the population is still developing its general ability to play IPD.

When tags are added to the mix, a small shift appears after generation 3,500. Throughout the entire run there are few significant differences between the run with tags and without, however, there is a trend in that the run with tags consistently shows the modern generation dominating the archaic generation in a smaller proportion of the 900 comparisons. This trend becomes more pronounced after generation 3,500. Despite this difference, the majority of the values are not significantly different between the run with tags and the run without tags as determined by the overlapping confidence intervals. Based on this data, it appears that tags have little to no effect on non-localized adaptation. This result is somewhat surprising as one would surmise that the addition of tags would make the task more complex. This added complexity should give the agents a larger strategy space to pursue, which should lead for more opportunity for non-localized adaptation.

The Survey of Tag and Tag Array Mutation Levels   One potential reason for the minimal effect of tags could have been their relatively low mutation level in the initial experiment. To address this possibility, we performed a 3 × 3 survey of tag and tag array mutation levels. With this type of an approach it is possible to see the effect of increasing tag related mutation levels as well as any interaction that may result from different relative tag and tag array mutation levels.

An initial look at the data in the table of Figure 3.29 shows that for most combinations there was little effect on non-localized adaptation. One condition, lowmed, showed a consistently higher proportion of dominance, however, many of the data points are not significant. In terms of significance, one condition that jumps out is the medhig condition. For this set of mutation rates, the proportion of dominance by the modern generation is consistently and significantly lower than the results for the data set without tags. The medmed condition gives a similar performance, however, there is a stretch without significance around generation 4,500. Despite these individual results, there is no apparent trend based on relative combinations of tag and tag array mutation levels. Regardless of the combination, there is little difference between populations with tags and without tags.

In light of these results, we decided to take a look at tag usage as these populations evolved. Our initial expectation was that most of the tag space would be occupied by agents. The data show the opposite. For all conditions the agents tend to all have the same tag. This means that there is only selective pressure for the portion of the FSM that deals with the current tag. Eventually, an agent's tag will mutate and that new agent will take over the world. It may well be able to do this because the other agents have no strategy for dealing with this new tag. The result is a population locked into a single tag value with punctuated changes in the tag value when a mutated agent is able to fill the open niche by mutating to a tag value that the rest of the population has no defense against. This invasion and takeover may slow the rate of non-localized adaptation to the nIPD task by creating periods of time when the selective pressure is reduced by a lack of strong "defensive" strategies to this new label.

When looking at the mutation levels that produced interesting results in the 3×3 survey, all of these conditions exhibited this tag cluster phenomenon. That is, most of the time the entire population had the same tag value. Occasionally, the population would exhibit multiple tag values during the time when an opportunistic agent was taking over the world, then the population would settle back down to predominantly one tag. Figure 3.4 shows some typical populations. In this graph, the x-axis represents the generation, the y-axis represents a simple measure of the amount of the tag space occupied. To calculate this metric, the total number of agents occupying each tag value is divided by number of agents that occupy the dominant tag value. These values are then summed to arrive at a single value for every 100th generation. For example, if there were 4 possible tag values and the distribution of 100 agents within the tag space was: $1:0, 2:50; 3:50, 4:0$. The measure of tag space utilization would be $(0/50 + 50/50 + 50/50 + 0/50) = 2$. With this metric, values close to 1 represent a population with very little diversity of tags, a value of 20 would represent a population evenly occupying the entire tag space.

The examples in Figure 3.31 are taken from the medmed mutation level. In this figure, Population A demonstrates a population that consistently occupies only one tag value at a time.

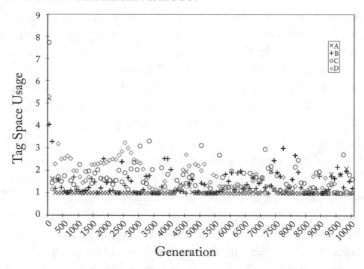

Figure 3.31: Graph depicting typical tag space usage for populations with tags.

When a new tag value takes over, the transition is fast and the population quickly returns to one tag value. Population B is a population that has a consistent, but extremely small level, of other populations but again one tag is typically dominant. In this case opportunistic agents seem to occur more frequently. These opportunistic agents prevent the population from being able to settle down. Despite this, there is still one dominant tag throughout the evolutionary history of the population. Population C is a variation on the same theme as population B. The only difference is that there are a couple of periods when there are multiple opportunistic agents at the same time, although these periods do not last for a long time and the population quickly settles down. Population D is an interesting example where there are fairly long periods of time where more than one tag is being used. This happens early in the population, but eventually a single tag prevails. Although interesting, the behavior of population D is quite rare and even when it does exist, it eventually comes back to the tag clustering problem.

This kind of a pattern can be seen in nature. The immune system utilizes surface proteins to identify and attack foreign organisms. Some organisms will change their surface proteins so that they can evade the immune response. One specific example of this can be seen in the malaria parasite. Strains of malaria, i.e., *Plasmodium falciparum*, have a set of genes called the var gene family that encode surface proteins. This particular parasite eludes the immune system by changing var gene expression which changes the proteins being displayed to the immune system, allowing it to stay ahead of the immune response [77]. In the population with tags it is easy to imagine that a similar phenomenon is occurring. When the entire population has the same tag, there is selective pressure to develop an response to the prevalent tag. Eventually a mutation develops where the tag of an agent is changed. Suddenly the entire population has no defense against this new tag. The agent then takes over the world by virtue of the population having no adequate defense.

Overall Mutation Rate   As mentioned before, the introduction of tag-related mutations with no corresponding increase in the overall mutation rate is similar to reducing the rate of FSM-based mutations. We performed a several data runs to isolate the effect of this decreased FSM mutation rate on the populations. Figure 3.30 shows these results.

When looking at the graph we can see that increasing the mutation rate by a factor of 10 had no significant effect on non-localized adaptation in the population. Interestingly, lowering the mutation rate by a factor of ten only had an effect on the initial rate of non-local adaptation, but it eventually merges with the other two data sets. This data seems to indicate that the overall mutation rate primarily has an effect in the early phase of an evolving population. After about 3,000 generations the effect of the overall mutation rate seems to fall off.

3.4.4 CONCLUSIONS

Non-localized adaptation seems to be a fairly resilient quality of an evolving population. As one might expect, populations acquire non-local adaptive features rapidly while they are evolutionary young. This study suggests that populations are still developing non-local adaptive features at later evolutionary times, as late as 9,500 generations. The addition of tags to the IPD game has little effect on the acquisition of adaptive features. One trend that can be seen in the data is that it becomes harder and harder to detect non-localized adaptation as the populations evolve. This result is not unexpected as it is easier for new mutations to enhance IPD play in the young initial populations rather than in the more evolved populations. The result is that a mutation is more likely to increase fitness early in evolution rather than later. Changes in the overall rate of mutation exhibit most of their effect early in the evolutionary process. As the population becomes more mature, the data tends to converge. A surprise from the study is the use of tag space. Although we expected the tag space to be uniformly used, instead the space is used in a virus like manner.

Throughout this book we have used multiple z-tests to look for significant differences between experimental populations. When making these comparisons it is typical to use a post-hoc correction such as the Bonferonni correction. These corrections are known to be overly conservative and, in this case, are so conservative that the 99% CI would be excessively wide. In future work we plan on making use of more sophisticated statistics in analysis of the data.

In this section we have taken a close look at the effect of mutation rates on non-local adaptation early in evolution. It would be interesting to determine the effect of these mutation rates on the length of time that a population exhibits non-local adaptation. One would imagine that a more complex task or a lower mutation rate would prolong the amount of time that a population was evolving non-local adaptive features.

The section examines a number of qualities—geography, tags, and changes in the character of mutation, that would be good to examine individually in the future. These factors are not the main focus of the book and so this section serves to at least raise the issue of these factors. In particular, the epochal character of the experiments in the first three seconds demonstrates that *time of evolution* is a critical variable.

CHAPTER 4

Does All This Happen Outside of Prisoner's Dilemma?

This chapter delves into examples of games other than Prisoner's Dilemma in which representation and resource levels have an impact on simulation. The novel game, Coordination Prisoner's Dilemma, is introduced as a partner to Rock-Paper-Scissors to create a continuous space of games. While the goal of the investigation of this continuous games space is to determine if zero sum games are harder as agent-training problems, the issues of agent resources were also investigated and in fact the paper this section is based on is where the issue of agent resources was first identified as an issue of concern.

Section 4.2 generalized the classic game Divide-the-Dollar to be able to model a wider variety of situations. This section is based on joint work of Daniel Ashlock and Garrison Greenwood. In this section a single sophisticated representation is derived and compared with results from replicator dynamics, demonstrating that the use of a less-than-basic representation permits the agents to behave in a more complex manner.

Section 4.3 is based on joint work of Garrison Greenwood and Daniel Ashlock. It is included to put forward his case for using the Snowdrift game, rather than Prisoner's Dilemma, as the workhorse for agent training research. The paper [107] addresses the issue of agent resources by varying the agent's memory depth.

4.1 COORDINATION PRISONER'S DILEMMA, ROCK-PAPER-SCISSORS, AND MORPHS

A common goal of computational intelligence in games is to generate agents that play a game well. In this section we are going to use very simple games to check the performance of co-evolution on a set of games that vary from zero-sum to far from zero-sum. The experiments in this section are intended to check if co-evolution is less effective at finding skilled agents for zero-sum games.

One problem that can occur in co-evolution is that the strategies cycle in successive generations instead of progressing [62]. A cause of this is the presence of intransitive relationships between strategies. This is the situation in which, given three strategies, A, B, and C, no one strategy beats both the others. A beats B, and B beats C, but C beats A. Intransitivity has been identified as a major issue in co-evolution (e.g., [72]), and was up to a point in time considered an unavoidable by-product of co-evolutionary dynamics, along the same lines that getting stuck in local minima was one of the pitfalls of artificial evolution. The identification of these issues lead to a series of algorithms, whose main aim was to stop co-evolution from reinventing solutions while not progressing toward a global goal.

It might be worth noting that most of the solutions proposed relied on defining a clear solution concept, i.e., when or under what circumstances a strategy is considered "better" than another, alongside a suitable strategy representation. If the solution concept and the strategy space are designed properly, then one can hope for a clear ranking between strategies. Unfortunately, it is often the case that such schemes produce algorithms that are slow to converge. Thus, in this section, we will not try to explicitly account for intransitivities in game strategies, but rather use simple co-evolution and try to understand under what circumstances one can see progress by only using local fitness information (i.e., the score of an agent against another).

Figure 4.1 shows all possible competitive arrangements among three strategies. One quarter of these arrangements are transitive. If co-evolution locates more effective agents, in effect finding agents that beat ever-higher fractions of all other strategies, then the fraction of transitive triples in the population should drop; failure to locate better agents will leave the fraction-of-transitive-triples statistic near 0.25 (and thus some kind of probabilistic strategy mixing might be required for an optimal agent [88]). When determining who the victor is in a pair within a triangle, we always orient ties in a consistent direction so that three way ties are measured as being transitive. Transitive fractions above 0.25 thus probably indicate the presence of common strategies adopted by many agents, which leads to numerous three-way ties. The goal of the research presented here is to test the following hypotheses.

1. Evolved populations of agents for zero-sum games are more likely than those for non-zero-sum games to have transitive triples of players.

2. The rate at which evolving populations are taken over by new agent types is higher when the agents are evolving to play a zero-sum game than a non-zero-sum game.

3. Being zero-sum is a singular quality in the space of games in the sense that zero-sum games are very different even from games that are close to being zero-sum.

One consequence of these hypotheses is that it is easier for co-evolution of players to engage in a cyclic and unproductive exploration of the strategy space for a zero-sum game than a non-zero-sum game.

The technique we use to explore these hypothesis is to evolve agents to play a variety of games chosen from games with payoff matrices from an α-parametrized family

$$M_\alpha = (1 - \alpha)M_1 + \alpha M_2, \tag{4.1}$$

where M_1 is zero-sum and M_2 is not.

The remainder of this section is structured as follows. Section 4.1.1 gives the experimental design including the agent representation, evolutionary algorithm structure, and describes the

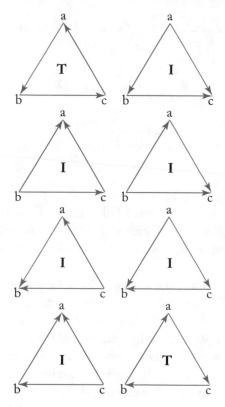

Figure 4.1: Depicted are the eight possible patterns of victory and loss for three strategies. Arrows point from the winner to the loser. Two of the four triangles are transitive, marked with a **T**, six are intransitive, marked with an **I**.

various analysis tools used. Section 4.2.3 gives the results and discusses them. Conclusions and next steps appear in Section 4.1.3.

4.1.1   DESIGN OF EXPERIMENTS

The endpoints of the space of games used in this section are Rock-Paper-Scissors and Coordination Prisoner's Dilemma using the payoff matrices shown in Figure 4.2. The payoff matrix for RPS was chosen to have the same total absolute value as the CPD matrix. The CPD game is introduced in this paper and requires some explanation.

The space of games defined by Equation (4.1) requires two endpoints, one of which is zero-sum. The M1 matrix at the zero-sum end is a RPS matrix. RPS was chosen because it is the best-known simple zero-sum game. It is a three-move, two-player game. Two-move zero-sum games are uninteresting, as one move has to beat the other, and ties have to result in a score

.	R	P	S
R	0	-4	4
P	4	0	-4
S	-4	4	0

.	D	C_a	C_b
D	2	6	6
C_a	1	4	0
C_b	1	0	4

Figure 4.2: The payoff matrices for RPS (left) and CPD (right) used in this paper. The score given is the score the player who plays the move indexed by the row receives when played against the move indexed by the column.

of zero. Our first thought was to use Prisoner's Dilemma, an exhaustively studied non-zero-sum game, as the other endpoint. However, PD has only has two moves. For compatibility with RPS, we developed CPD, a three move non-zero-sum game that combines PD with a coordination game.

CPD has three moves, defect (D), and two distinct types of cooperation, C_a and C_b. The subgames for the move sets D, C_a and D, C_b are identical and both obey the defining inequalities for Prisoner's Dilemma [39]. The subgame for the move set C_a, C_b is a coordination game in which making the same move yields a high payout and making different moves yields a poor one. Considering the whole game, we have chosen to make mutual defection superior to failure to coordinate types of cooperation. The game is intended to model a situation similar to a large corporation or institution where agents engage serially in a large number of small joint tasks. In this environment the usual choices of cooperation and defection are nuanced to include competent, knowledgeable cooperation and well-intentioned, but incompetent, cooperation.

Agent Representation   The agent representations used in this section are 8- and 64-state finite state machines (FSMs) with actions associated with transitions between states (Mealy machines). FSMs were chosen as the agent representation because they are capable of encoding a broad variety of strategies and because several analysis techniques have been developed for the analysis of finite state machines that do not translate smoothly to other agent types defined in Chapter 2. State transitions within the finite state machines are driven by the opponent's last action. Access to state information permits the machine to condition its play on several of its opponent's previous moves. The machines are stored as linear chromosomes listing the states. Each state includes actions and transitions for each of the three possible opponent actions. The initial action for a given machine are stored with and undergoes crossover with the first state in this linear chromosome. The initial state is always the first state stored in the chromosome.

Two variation operators are employed: a binary variation (crossover) operator and a unary variation (mutation) operator. The binary variation operator used is two-point crossover on the list of states. Crossover treats states as atomic objects. The mutation operator changes a single state transition 40% of the time, the initial action 10% of the time, or an action associated with a transition 50% of the time. The unary variation operator replaces the current value of whatever it is changing with a valid value selected uniformly at random. The object to change is chosen uniformly at random from those available. The use of two different numbers of states is intended to draw attention to the impact of the number of states, but examining this parameter in greater detail is beyond the scope of this initial paper. Eight states is based on past research on Prisoner's Dilemma, enough to represent effective strategies, while 64 represents a generous excess.

The Evolutionary Algorithm   The evolutionary algorithm used here operates on a population of 36 agents, a number chosen for compatibility with previous studies on Prisoner's Dilemma in Chapter 2 and [26]. Agent fitness is assessed with a round-robin tournament in which each pair of players engage in 150 rounds of the game under study. Reproduction within the algorithm is elitist with an elite of the 24 highest scoring strategies, another choice that maintains consistency with past studies. When constructing the elite, ties are broken uniformly at random. Twelve pairs of parents are picked by fitness-proportional selection with replacement on the elite. Parents are copied, and the copies are subjected to crossover and mutation. A single mutation is performed on each new agent.

Two types of simulations were performed. The short simulations were run for 250 generations for a given agent type and choice of game and summary statistics on the fitness, fraction of actions of each type, and agent complexity (described below) were saved. Each short simulation contains 30 independent replicates and the final population is also saved. The long simulations are run for 4,000 generations and the most fit agent in generations 250, 500, 1,000, 2,000, and 4,000 is saved for analysis. These are used for the study of non-localized adaptation, define in Section 3.4. As before, the sampling intervals are called *epochs*. Each long simulation contains 1,000 independent replicates yielding a pool of five sets of 1,000 agents for analysis and comparison. The long simulations are also run for each agent type and choice of game.

The following sets of parameters are used for both short and long simulations. For both 8- and 64-state agents we use $\alpha = 0, \frac{1}{100}, \frac{1}{2}, \frac{8}{13}$ and 1. The value $\alpha = \frac{8}{13}$ is chosen because it represents a change in the meaning of the moves. In RPS, R beats S, and P beats R. In CPD, D (which corresponds to R) beats both C_a (which corresponds to P) and C_b (which corresponds to S). So, as the value of α changes from zero to one, P goes from beating R to being beaten by it. $\alpha = \frac{8}{13}$ is the point at which they tie. The other values of α are zero-sum, just barely not zero-sum, half-way, and as far from zero-sum as possible.

Analysis Techniques Used   One of the analysis techniques used in this section is transitivity or, more precisely, the fraction of agent triples that are transitive in the sense given in the introduction. This measure was first proposed in a co-evolutionary setting in [97], and is an application of ideas in [60]. This measurement is termed the *transitivity of an agent set*. If transitivity is computed over all triples for a given set of agents, then the measure is potentially brittle because the reversal of the direction of an arrow for a single pair of agents can cause a large change in the value. We avoid this problem by first generating a large number of agents and then sampling sets of independent triples.

For each agent representation, evolutionary epoch, and payoff matrix we use a set of 1,000 agents from independent evolutionary runs. These agents are shuffled into a random order and the 333 disjoint triples of index $3x$, $3x + 1$, $3x + 2$ (one agent is left over) are used to estimate the fraction of transitive triples. 400 orderings are sampled to capture the variability of the estimate and compute a 95% confidence interval on the transitivity estimate. This sampling covers $133, 200$ of the 167 million triples of agents in a 1,000-agent sample.

Complexity Measures   We use two complexity estimates, the sandpile and diffusion character estimates. The sandpile measure addresses the asymptotic behavior of a finite state machine while the diffusion character measure addresses overall complexity including transient states in the FSM. Both these complexity measures are similar to Markov-chain entropies. Both measures treat the states and transitions of the FSM as a directed graph. The sandpile measure places 1.0 unit of sand on the initial state and then iteratively divides the sand among neighbors for each state where sand is present. The process is run for a large number of steps and, once the distribution of sand is stable or periodic, the Shannon entropy

$$E = -\sum_{p \neq 0} p \log_2(p), \qquad (4.2)$$

where p is the amount of sand in each state is used to characterize the complexity of the distribution. Since the maximum entropy is different for 8- and 64-state agents, we normalize to the fraction of maximum possible entropy.

The diffusion character estimate is similar but, has two differences. Rather than conserved sand, it uses gas with an additional unit of gas injected into the initial state in each time step. Each state absorbs 50% of the gas present in each iteration as well. As with the sand, the process is run long enough for the gas to reach its asymptotic distribution. The amounts of gas are then divided by the total gas to yield an empirical probability distribution whose Shannon entropy is the diffusion character complexity estimate.

This work is the first to use diffusion characters on the digraph of an FSM, so some explanation of how it is applied is required. Each state in a FSM has a transition for every possible move, three for the game-playing agents used here. It follows that when each transition is considered as a separate edge the out-degree of every vertex in the digraph of a FSM is equal to the size of the input alphabet, three for the games used in this section. This means that the adjacency matrix of the digraph of a FSM can be column normalized by dividing the (i, j)th entry in the adjacency matrix by the out-degree of the jth vertex. As such the digraph of a FSM can be analyzed with diffusion character matrices.

The diffusion character matrix of a graph is a transformation of the adjacency matrix of a graph that emphasizes the behavior of random walks in the graph. Diffusion character matrices first appeared in [24] where some of their mathematical properties were investigated, and they have subsequently been used as a method of defining pseudo-metric distances between graphs in [23, 49]. A diffusion character matrix is a parametrized transformation of a column stochastic matrix M defined as follows: for $0 < \gamma < 1$, the diffusion character matrix of the column stochastic matrix $M \in \mathcal{R}^{n \times n}$ is the matrix $(I_n - \gamma M)^{-1}$ which is equivalent to the matrix $\sum_{k=0}^{\infty} (\gamma M)^k$. The normalized diffusion character matrix of M is the matrix $(1 - \gamma)(I_n - \gamma M)^{-1}$ which is column stochastic.

When M corresponds to a column normalized adjacency matrix then $[M^k]_{i,j}$ is the probability that a random walk of length k starting at vertex j ends at vertex i. Thus, the (i, j)th entry of a normalized diffusion character matrix may be viewed as the geometrically weighted probability that a random walk starting at vertex j ends at vertex i, and hence the Shannon entropy of a column of the normalized diffusion character matrix may be viewed as a scalar summary of this probability distribution. For our diffusion character complexity estimate we are using the entropy of the column associated with the starting state of our agents.

Non-Localized Adaptation   This section uses a slightly different measure of non-localized adaptation from that give in Section 3.4. Informally, non-localized adaptation (NLA) is the degree to which additional evolution of a population of co-evolving agents improves their competitive ability. Non-localized adaptation was first explored in [25] where positive results were obtained for Prisoner's Dilemma playing agents. In [9], the effect was observed in virtual robots competing to paint the floor red or blue. Examples of non-local adaptation were found in simulations of ecologies in [6, 56]. Additional research on NLA in Prisoner's Dilemma appears in [19]. In the latter of these two studies it was found the NLA can break down after very long periods of evolution. The work presented in this section is the first to apply NLA to a simultaneous two-player game other than Prisoner's Dilemma.

NLA is estimated as the probability, with 10,000 pairs of agents sampled, that an agent from an earlier epoch will beat one from a later epoch. Numbers significantly below one-half are thus evidence for non-local adaptation. Numbers significantly above one half are evidence of retrograde non-local adaptation.

4.1.2   RESULTS AND DISCUSSION

The primary assessment used for the work presented in this section is transitivity. The results of the transitivity study are shown in Figure 4.3. We note that the deviation from the random expectation of one-quarter for agents playing the zero-sum game is either not statistically significant or, for the one significant difference, is quite small. All the non-zero-sum games exhibited statistically significant differences from expected random transitivity. The pattern of these differences, however, possesses a number of unexpected features.

The most unexpected feature is the significantly above expected transitivity found in the 8-state agents for $\alpha = \frac{1}{2}$ and $\frac{8}{13}$. The effect changes significantly across epochs and is very different for the two α values where it occurs Our tentative explanation is that the effect arises from an enhanced fraction of ties, since if all three agents in a group tie one another, then the triple is

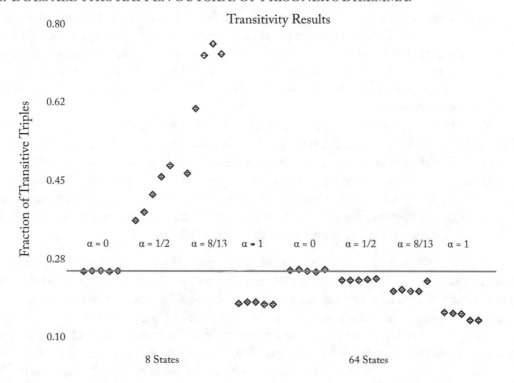

Figure 4.3: Shown are 95% confidence intervals on the fraction of transitive triples for 8- and 64-state machines and each payoff matrix for all five epochs. The epochs are the groups of five with age of the agents increasing from left to right. The gray line marks the random expectation of 1/4 for transitive triples.

counted as transitive. This could either be the result of many identical strategies that tie in self-play in our group of 1,000 or the result of the evolution of strategies that tie in play with many other strategies. Future work will redo the analysis keeping track of ties separately to confirm this interpretation.

Supporting these observations is the startling difference between the 8- and 64-state agents with the same α value. The space of 64-state strategies is much larger and so may contain more good strategies and take longer to converge to good strategies. This behavioral disjunction between 8- and 64-state agents is itself an unexpected result and the one that motivated the work reported in Chapter 3 on the impact of change resources available to agents. The transitivity results are significantly different for agents with different numbers of states for all tested α values except 0. This suggests that changing the number of states an FSM can qualitatively change its nature as a representation for game playing agents.

Although the transitivity results do not support the original hypothesis that zero-sum games will have more transitive triples than non-zero-sum games, they do provide evidence that co-evolution is not as effective for locating agents for a zero-sum game as a non-zero-sum game. The lack of a significant deviation from random expectation of transitivity is evidence that evolution is cycling through the same strategies and so not doing useful work. In all of the non-zero-sum games, the transitivity results provide clear evidence that evolution performed nontrivial work and acted to selected populations that lie in a subset of the search space.

Figure 4.4 shows the way the use of the three actions evolves in four example populations. The first column shows that RPS ($\alpha = 0.0$) has a high rate of succession. As an agent type becomes dominant, it unavoidably becomes an exploitable target because of the zero-sum nature of the game. A simple permutation of the three moves yields a new strategy that can beat an old one. This explains the rapid waves of change. CPD, $\alpha = 1.0$ shown in the rightmost column, on the other hand, is able to sustain dominant strategies or stable mixtures of strategies for hundreds of generations. When a blue or a green line is uppermost, the population is mostly cooperating. The figure in the second row demonstrates that both sorts of cooperation can occur in a stable population. The figure in the last row demonstrates cycling between a mostly cooperative population and a mostly defecting population. The middle game, when $\alpha = 0.5$, shows a variety of behaviors only a few of which are shown in the figure. The longest stable periods occur in some replicates for this game, but unstable evolutionary tracks like the one in the last row also occur. Figure 4.4 demonstrates that the games studied here yield both diverse behaviors and substantially different behaviors from one another.

Table 4.1 summarizes the complexity results. The most obvious result is that 64-state agents have higher complexities than 8-state agents. This is true of random 8-state and 64-state agents as well and is because the expected fraction of unused states in 8-state agents is much larger. Missing 1 of 8 states is far more likely than missing 8 of 64. This highlights the fact that the 8-state agents have a far smaller budget of states and that this smaller budget has an impact. To provide scale on the entries of Table 4.1 the measures were computed for 1,000 randomly generated agents. For 8-state agents the random expectations are estimated to be sandpile 0.8676 ± 0.0128 and diffusion 0.8822 ± 0.0083. The 64-state agents have random expectations estimated at 0.9371 ± 0.0015 and 0.9181 ± 0.0018 respectively.

In all cases, the complexity trends downward as we move the payoff matrix away from RPS until we reach $\alpha = 8/13$. For 8-states, it rises for CPD; for 64-states, complexity for the $\alpha = 8/13$ game and CPD are similar. The estimated complexities, using both measures, are close to the random expectation for RPS and smaller for $\alpha > 0$. This is further evidence that co-evolution is performing nontrivial work and selecting populations in a subset of the search space.

The last observation has to do with the difference between the two complexity estimates. The sandpile estimate measures the asymptotic character of the strategy; the diffusion character estimate measures overall complexity. This means that comparing the two gives information

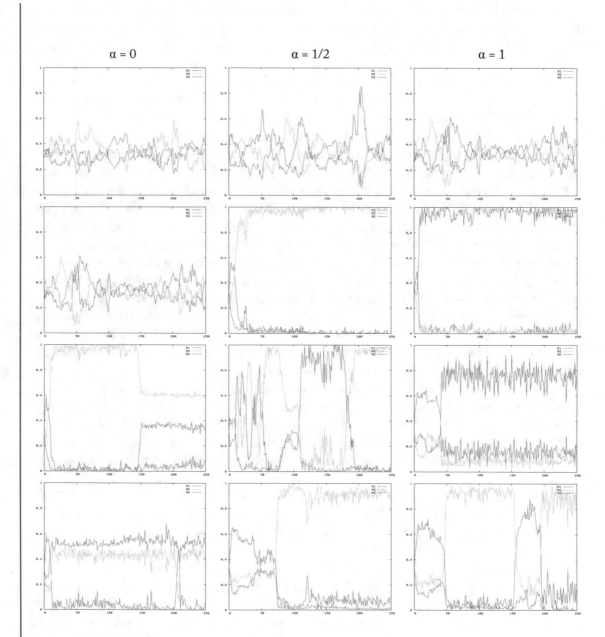

Figure 4.4: The graphs show the fraction of actions used by agents across an entire population, over the course of evolution, for four sample replicates for each of $\alpha = 0$, $\frac{1}{2}$, and 1.

Table 4.1: Tabulated are 95% confidence intervals on the sand-pile and diffusion-character complexity measures of groups of 1,000 agents from independent evolutionary runs for all the parameters tested as well as the differences between the complexity measures

Epoch	α	Sand-pile	Diffusion Character	Difference
8 States				
250	0	0.89339±0.00645	0.89466±0.00616	0.0013
500	0	0.88817±0.00787	0.89197±0.00689	0.0038
1000	0	0.89643±0.00887	0.89995±0.00647	0.0035
2000	0	0.89199±0.00829	0.89463±0.00747	0.0026
4000	0	0.89397±0.00699	0.89504±0.00629	0.0011
250	5/13	0.84364±0.01264	0.84840±0.00966	0.0134
500	5/13	0.82670±0.01474	0.83322±0.01176	0.0009
1000	5/13	0.82038±0.01587	0.82934±0.01280	0.0010
2000	5/13	0.82757±0.01341	0.82883±0.01099	-0.0059
4000	5/13	0.83634±0.01217	0.83459±0.01138	-0.0166
250	1/2	0.80147±0.01709	0.81484±0.01198	0.0176
500	1/2	0.78131±0.01766	0.78222±0.01536	0.0185
1000	1/2	0.79550±0.01804	0.79652±0.01432	0.0155
2000	1/2	0.78231±0.01751	0.77645±0.01497	-0.0037
4000	1/2	0.79413±0.01732	0.77750±0.01556	-0.0020
250	1	0.75677±0.02335	0.77804±0.02000	0.0213
500	1	0.74389±0.02286	0.76503±0.01864	0.0211
1000	1	0.73597±0.02397	0.76729±0.01873	0.0313
2000	1	0.75432±0.02106	0.77800±0.01673	0.0237
4000	1	0.73368±0.02320	0.76147±0.01839	0.0278
64 States				
250	0	0.93687±0.00145	0.91740±0.00165	-0.0195
500	0	0.93617±0.00163	0.91699±0.00171	-0.0192
1000	0	0.93573±0.00143	0.91555±0.00169	-0.0202
2000	0	0.93548±0.00153	0.91562±0.00166	-0.0199
4000	0	0.93544±0.00151	0.91558±0.00193	-0.0199
250	5/13	0.93109±0.00176	0.91082±0.00226	-0.0153
500	5/13	0.92796±0.00173	0.90637±0.00226	-0.0170
1000	5/13	0.92704±0.00193	0.90686±0.00209	-0.0210
2000	5/13	0.92631±0.00193	0.90384±0.00229	-0.0217
4000	5/13	0.92872±0.00171	0.90661±0.00210	-0.0251
250	1/2	0.92438±0.00660	0.90904±0.00201	-0.0161
500	1/2	0.92245±0.00491	0.90545±0.00207	-0.0179
1000	1/2	0.92194±0.00221	0.90096±0.00256	-0.0165
2000	1/2	0.92043±0.00237	0.89872±0.00275	-0.0124
4000	1/2	0.92334±0.00213	0.89827±0.00266	-0.0227
250	1	0.92753±0.00190	0.90881±0.00198	-0.0187
500	1	0.92109±0.00498	0.90314±0.00233	-0.0180
1000	1	0.91375±0.00674	0.89777±0.00271	-0.0160
2000	1	0.90098±0.01114	0.89271±0.00269	-0.0083
4000	1	0.90703±0.00809	0.89292±0.00265	-0.141

about transient states. Transient states are states from which it is possible to permanently depart. These are useful for kin recognition. Prisoner's Dilemma strategies have been shown to develop kin recognition as they co-evolve. For RPS there is no benefit to kin recognition, because kin are strategies that produce ties that score zero. Instead, the primary evolutionary dynamic in RPS is replacement of the currently most successful agent type with its natural successor.

The last column in Table 4.1 shows the diffusion character estimate minus the sandpile estimate. Values near zero suggest transient states are not being used. Values larger than zero suggest transient states are more complex than recurrent states; values smaller than zero suggest recurrent states are more complex. Note that the 8-state values are mostly positive, while the 64-state values are negative. This is also true of random 8-state and 64-state FSMs, with the difference being 0.0146 for 8-state random machines and −0.0190 for 64-state random machines. This is caused by the fact that a single transient state is a greater proportion of an 8-state machine than a 64-state machine. Assuming that the number of states needed to recognize kin is the same for both 8-state machines and 64-state machines, 8-state machines face a trade-off between using their states for kin recognition and using them to develop more complex strategies that 64-state machines do not.

Figure 4.5 shows the difference in complexity values for 8-state machines shifted so that zero is the expectation for random machines, and Figure 4.6 shows the same thing for 64-state machines. Comparison of these figures makes it clear that transient states have a bigger impact on 8-state machines than on 64-state machines. Similar patterns occur in the two figures, however. Transient states have the largest impact on the evolution of CPD strategies, and negligible impact for RPS. The only within game difference with statistical significance is in the 8-state $\alpha = 1/2$ game. For this game, strategies in generation 250 have higher diffusion character entropy, suggesting complexity in the transient states, and strategies in generation 4,000 have higher sandpile entropy, suggesting complexity in the recurrent states.

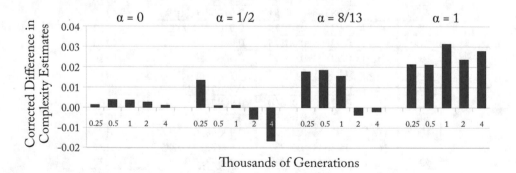

Figure 4.5: Difference in complexity measures (diffusion characters estimate minus sandpile estimate) for 8-state machines displayed so that zero is the difference expected in random machines.

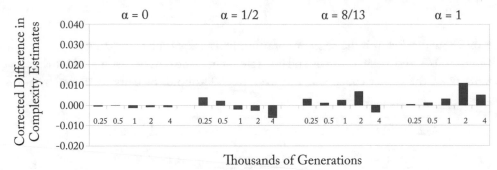

Figure 4.6: Difference in complexity measures (diffusion characters estimate minus sandpile estimate) for 64-state machines displayed so that zero is the difference expected in random machines.

Table 4.2 gives the NLA results. For RPS both sorts of agents exhibited small amounts of NLA, often not significant. In the other games, 8-state agents all exhibited significant NLA for all pairs of epochs tested with the strongest NLA appearing in the 8-state game with $\alpha = 8/13$. In sharp contrast, the 64-state agents exhibited either insignificant NLA or significant retrograde NLA. Retrograde NLA results from over-specialization to an agent's particular environment. This is additional evidence that there is a qualitative difference between the agent's representations with different numbers of states.

4.1.3   CONCLUSIONS

Our primary result is to support the conjecture that co-evolution finds zero-sum games a more difficult target than non-zero-sum games. The results of this research are quite narrow and should not necessarily be generalized to other simultaneous two-player games. The co-evolutionary behavior of other zero-sum games is discussed in [97] where an algorithm that minimizes cycling is presented and applied to the co-evolution of the zero-sum game of Othello.

We now evaluate our three hypotheses from the Introduction.

1. The bizarre results for 8-state agents on the intermediate game show that the first hypothesis was insufficiently nuanced. If we had claimed that zero-sum games stay closer to the random expectation of transitive triples we would have scored a success. Our failure lay in not imagining that we could get agents with a significantly enhanced fraction of transitive triples.

2. The second hypothesis, that succession of new types is more rapid for zero-sum games, is supported. Figure 4.4, the transitivity results, and the complexity results all support this conclusion in different ways. RPS populations cycle through use of different proportions of the three moves; they have numbers of transitive triples near random expectation; and

Table 4.2: Tabulated below is the degree of non-localized adaptation comparing all possible pairs of epochs for α=0, 1/2, and 1 using both 8- and 64-state agents. The number estimated is the probability that agents from the earlier epoch will win over 10,000 samples of pairs of agents from distinct epochs.

Epochs Compared	8 States	64 States
$\alpha = 0$ –RPS		
250 vs. 500	0.4675+/-0.0245	0.4956+/-0.0245
250 vs. 1000	0.4906+/-0.0245	0.4781+/-0.0245
250 vs. 2000	0.4625+/-0.0244	0.4956+/-0.0245
250 vs. 4000	0.4875+/-0.0245	0.4969+/-0.0245
500 vs. 1000	0.4831+/-0.0245	0.4731+/-0.0245
500 vs. 2000	0.4631+/-0.0244	0.4544+/-0.0244
500 vs. 4000	0.4781+/-0.0245	0.4763+/-0.0245
1000 vs. 2000	0.4731+/-0.0245	0.4738+/-0.0245
1000 vs. 4000	0.4756+/-0.0245	0.4944+/-0.0245
2000 vs. 4000	0.4900+/-0.0245	0.4781+/-0.0245
$\alpha = 1/2$		
250 vs. 500	0.2944+/-0.0223	0.4994+/-0.0245
250 vs. 1000	0.3144+/-0.0227	0.5275+/-0.0245
250 vs. 2000	0.2925+/-0.0223	0.5663+/-0.0243
250 vs. 4000	0.2644+/-0.0216	0.5719+/-0.0242
500 vs. 1000	0.3000+/-0.0225	0.5394+/-0.0244
500 vs. 2000	0.2788+/-0.0220	0.5356+/-0.0244
500 vs. 4000	0.2581+/-0.0214	0.5475+/-0.0244
1000 vs. 2000	0.2575+/-0.0214	0.5400+/-0.0244
1000 vs. 4000	0.2781+/-0.0220	0.5375+/-0.0244
2000 vs. 4000	0.2594+/-0.0215	0.5019+/-0.0245
$\alpha = 1$ –CPD		
250 vs. 500	0.3913+/-0.02391	0.4663+/-0.0244
250 vs. 1000	0.3963+/-0.02397	0.4800+/-0.0245
250 vs. 2000	0.4094+/-0.02409	0.5094+/-0.0245
250 vs. 4000	0.3644+/-0.02358	0.5781+/-0.0242
500 vs. 1000	0.3713+/-0.02367	0.4788+/-0.0245
500 vs. 2000	0.4081+/-0.02408	0.5169+/-0.0245
500 vs. 4000	0.3869+/-0.02386	0.5575+/-0.0243
1000 vs. 2000	0.3925+/-0.02393	0.5313+/-0.0245
1000 vs. 4000	0.3913+/-0.02391	0.5669+/-0.0242
2000 vs. 4000	0.3850+/-0.02384	0.5244+/-0.0245

their agents are more difficult to tell from random agents than those evolved for other payoff matrices.

3. We found no support for the hypothesis that zero-sum games form a singularity in game space. Results not shown for $\alpha = 0.01$ were never significantly different from those for $\alpha = 0$. When play is iterated for 150 rounds in fitness evaluation the transitivity of payoff matrices for small values of α yields the same sort of rapid succession of new types that was observed for $\alpha = 0$.

The qualitative difference between the agents with different numbers of states was an unexpected result in this work. This extends and agrees with earlier results presented in Chapter 2 on the importance of the choice of representation for game playing agents. The earlier study compared radically different representations such as FSMs, artificial neural nets, parse trees, and stochastic look-up tables. This paper demonstrates that even using the same type of representation with different parameters can affect the course of evolution. There were many nonlinear responses observed for the parameter α. Transitivity, complexity, and NLA were nonlinear in α.

A clear next step is to verify that other types of zero-sum games are more difficult for evolution to deal with than related non-zero-sum games. The startling results connected with changing the number of states in the finite state agent also merit examination. Any quantity that varies has local or global optima. If we choose competitive ability as our objective function it would be interesting to characterize what the most effective numbers of states are for agents for playing various simultaneous two- or three-player games. The CPD game was invented to be "far from" RPS. This work suggests that the game is interesting in its own right and merits additional study. This research fulfills an often repeated item of future work for several of the authors: moving beyond basic two-move Prisoner's Dilemma.

4.2   DIVIDE-THE-DOLLAR—A MORE COMPLEX GAME

In this section the game Divide-the-Dollar is generalized and two representations for it are contrasted—a replicator dynamic model that does not use agents and an agent-based model using finite state agents. Instead of the two or three moves available in the games presented thus far, in this book, Divide-the-Dollar has 101 distinct moves (0–100 pennies) and so requires greater representational finesse.

Divide-the-Dollar is a two-player simultaneous game invented by John Nash because its strategy space has an entire subspace of Nash equilibria. In this section we describe and explore a family of generalizations of Divide-the-Dollar with easily controlled properties. The game Divide-the-Dollar is a simplification of the bargaining game proposed by John Nash [87]. The game is a simultaneous two-player game in which each player makes a bid. If the bids total a dollar or less then each player receives their bid, otherwise they receive nothing. The space of moves for this game is the set of pairs of positive fractions of a dollar, with the pairs summing to no more than one yielding positive payoffs. A feature that makes this game interesting is that there is an entire subspace, pairs of move that sum to one dollar, that are all Nash equilibria.

A *Nash equilibrium* of a game is a set of moves, one per player, so that no player may improve their score by changing their move unilaterally. The Nash equilibria of a game are places where, in classical game theory, players are likely to arrive. Having an entire subspace of Nash equilibria means that player behavior is harder to predict. There is, *a priory*, no game-theoretic reason to prefer and even division of the dollar to a 90:10 split. Humans will see a 50:50 split as being fairer, but this is not a notion encoded in the definition of Nash equilibria.

In a pair of papers that used genetic programming to train agents to play Divide-the-Dollar [2, 27] it was found, first, that evolved agents favor near even splits when they are drawn from a single breeding population. The second result was that when we split the players into two breeding populations, with all play involving one agent from each population, that the tendency toward results a human would think more equitable decreased substantially. In essence the evolution of fairer solutions was a kinship effect [59]. The representation used in these two papers is quite complex and that complexity may have been partially responsible for the results.

In [1], a modification of Divide-the-Dollar that removes the problematic subspace of Nash equilibria is presented together with a n-player generalization of the game. In this section we study a larger generalization that permits any number of players and permits the researcher to design features into the game by selection of a critical set that defines the payoff space. This initial work examines only two-player versions of the game; it explores the ability to design generalizations in a simple manner.

Why Divide-the-Dollar?   A good deal of work has been done in evolving agents to play the IPD [28, 41, 46, 74, 95, 98]. This work revisits and generalizes the evolution of agents to play Divide-the-Dollar because many of the advances made in studying and understanding Prisoner's Dilemma can be applied to Divide-the-Dollar and its generalizations, but the games derived from Divide-the-Dollar are more complex and better suited to modeling many economic situations than Prisoner's Dilemma.

With two possible moves in each play, Prisoner's Dilemma yields the minimal potentially useful resolution in modeling economic behavior. Divide-the-Dollar has around a 100 moves, permitting incremental changes in the character of moves, and a far richer strategy space. In addition, the conflict in Divide-the-Dollar is materially different from that in Prisoner's Dilemma. While bidding so high as to disrupt a deal has the flavor of defection about it, greed and defection have different characters. In Prisoner's Dilemma the very high temptation payoff means that the best payoff is for successful greed. In Divide-the-Dollar, greed simply receives a penalty. This means that the temptation to defect is less and the dilemma—"I do better if I defect" is much weaker and more nuanced in Divide-the-Dollar.

One practical outcome of this difference is that knowledge of the number of rounds of play is not as important in Divide-the-Dollar. In the IPD, if a player knows the current round of play is the last, then there is no selfish motive to do anything but defect. There is no analogous motive to bid high in Divide-the-Dollar. When evolving agents to play IPD the number of rounds of play must either be randomized or large enough to defeat any given agent representation's ability to count the number of rounds. In Divide-the-Dollar there is no such need for random or large number or rounds of play.

4.2.1   GENERALIZE DIVIDE-THE-DOLLAR

If $S \subset \mathbb{R}^n$ then S can be used to specify an n-player generalization of Divide-the-Dollar with the following mechanic. Each player supplies a coordinate of a point in $\mathbb{R}^n$. If the resulting point is in the set S then each player gets the coordinate he supplied as his score, otherwise the players all receive zero. We will call this new game *Generalized Divide-the-Dollar* (GDD) for a set S.

In any game, the knowledge a player has of the rules of the game are an important part of the setup. This can range anywhere from forcing a player learn the rules as he goes along to giving a complete specification of the rules. In this game, exact knowledge of the set, together with the scoring rules, represents complete knowledge of the game. A reasonable partial state of knowledge is a specification of which of a players moves have a chance of scoring.

Definition 4.1   In a round of GDD the players are said to have **coordinated** if they jointly choose a point in the set.

Definition 4.2   Each player have control of one coordinate of a point. The **marginal support** for a player in a set S is the set of numbers that are coordinates points in S that are under the player's control. This set is also the projection of the set on the axis corresponding to the player.

In this section:

- At a minimum, each player is presumed to know the marginal support of the set in his own coordinate.

- A more information-rich version of the game permits the player exact knowledge of the set.

- In an iterated version of the game the player can receive feedback consisting of only their score or might be given the moves made by all players.

If a player knows the set and the other players moves under iteration we call this a *total information* version of the game. If the player knows only his own coordinates marginal support and the score obtained then this is the *personal information* version of the game. If the player knows the full set but only his on score, we call this the *semi-global* version of the game. If a player knows his own marginal support and the other players full move under iteration, we call this the *exploration* version of the game. The player can mine other player's moves to gain information about the game.

In this work, evolving finite state agents are used. The agents adapt to the outcomes via evolution and so learn appropriate responses based on feedback reporting one of three conditions: the agent received the higher score, the agent did not received the higher score, or the agents failed to coordinate. This work thus uses an exploration version of the game.

The set S given in the left half of the first panel of Figure 4.7 specifies the classic Divide-the-Dollar game. The graphs shown in this figure give the asymptotic distributions of bis under

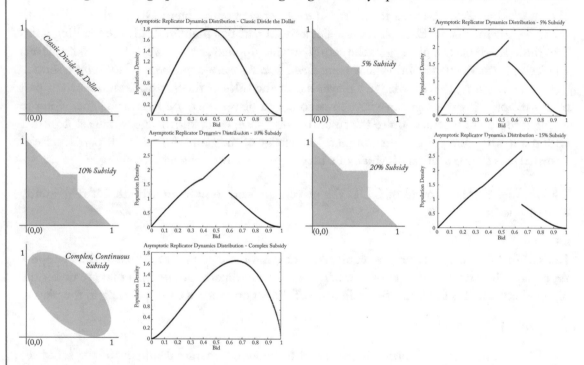

Figure 4.7: The five sets used for exploring the properties of the Generalized Divide-the-Dollar game. The top set is the classic Divide-the-Dollar; the middle three are the modifications with a simple central subsidy for being within 5%, 10%, and 15% of a bid of 50 cents; the last is the complex central subsidy modeled with a rotated ellipse. Next to each set is a plot of the asymptotic distribution of bids that arise under replicator dynamics.

replicator dynamics. For a given set we adopt a uniform initial distribution of bids and then iteratively update it according to the following replicator dynamics scheme. If D_0 is the initial uniform distribution then,

$$\widehat{D_{n+1}}(v) = v \cdot \int_{S \,:\, v \text{ scores}} dA \qquad (4.3)$$

$$D_{n+1}(v) = \widehat{D_{n+1}}(v) / \int_S \widehat{D_{n+1}} \cdot dA, \qquad (4.4)$$

where dA is the differential of area, volume, or hyper-volume, as appropriate. While we exclude the details here, the updating procedure is a contraction map and so the series of distributions in which the distribution has no interior zeros converges to a unique fixed point. Examples of fixed asymptotic distributions for the sets used in this work appear in Figure 4.7.

This experiment uses five sets, shown in Figure 4.7. The first of these sets yields the classic Divide-the-Dollar game. These sets are all subsets of the unit square and so have the constraint $0 \leq x, y \leq 1$. In standard mathematical notation, these sets are:

1. $S_1 = \{(x, y) : x + y < 1\}$,

2. $S_2 = \{(x, y) : x + y < 1 \text{ or}$
   $(|x - 0.5| < 0.05 \text{ and } |y - 0.5| < 0.05)\}$,

3. $S_3 = \{(x, y) : x + y < 1 \text{ or}$
   $(|x - 0.5| < 0.1 \text{ and } |y - 0.5| < 0.1)\}$,

4. $S_4 = \{(x, y) : x + y < 1 \text{ or}$
   $(|x - 0.5| < 0.15 \text{ and } |y - 0.5| < 0.15)\}$, and

5. $S_5 = \{(x, y) : (x - 0.5)^2 + (y - 0.5)^2 + 1.2(x - 0.5)(y - 0.5) < 0.16\}$.

The first of these sets reproduces the classical Divide-the-Dollar game. Sets 2–4 represent a type of subsidy intended to encourage play within a range near the 50/50 divide. The area of the set above the line $x + y$ represents the available money from external subsides. Players that place bids within a range around 50/50 receive extra profit. The final set represents a subsidy that is smooth, avoiding the discontinuities that appear in the asymptotic replicator dynamic plots for sets 2–4. The elliptical set encourages bids near the center, albeit less harshly than the other subsidies. It also penalizes bids that leave too much of the benefit available in a deal on the table. The area between the ellipse and the origin is the area in which this set penalizes play.

Since the nonstandard sets are modeling subsidies, it is important to figure out the cost of these subsidies. Since the units of area are squared dollars, the cost (per bid) of a subsidy is reasonable to estimate as the square root of the area in the set above the line $x + y = 1$. The subsidies are the *maximum cost per bid* if the agents optimally exploit the subsidy. These values are given in Table 4.3.

Nash Equilibria for GDD   The Nash equilibria for GDD are easy to compute once a simple definition has been made.

Definition 4.3   A pair (a, b) of bids that yield non-zero payoffs **dominate** another such pair (c, d) if $a > c$ and $b > d$. A pair is **non-dominated** if no pair dominates it.

It follows directly from the definition of Nash equilibria that the Nash equilibria for a given instance of GDD are the non-dominated pairs of bids.

Table 4.3: The area of each of the sets and the cost of the subsidies they encode

Set	Area	Subsidy Cost
S_1	0.5000	0
S_2	0.5005	0.0707
S_3	0.5020	0.1414
S_4	0.5045	0.2121
S_5	0.3142	0.5605

4.2.2 DESIGN OF EXPERIMENTS

In this experiment we examine five different versions of GDD based on the sets given in Figure 4.7. We will compare the naive replicator dynamic distribution of bids for each set with the distribution of bids that arise in the last generation of evolutionary algorithms training agents to play GDD. The ability of agents to learn to coordinate bids are also tracked. The sets used have different areas—and larger areas would make randomly coordinating bids easier. Table 4.3 gives the area of each of the sets and the cost of its subsidy, as defined in the preceding section. Notice that the complex subsidy has a very high relative cost.

The Agent Representation   The agents used in this experiment are finite state Moore machines that generate point coordinates as their output. Moore machines associate the output values they emit with their states, as opposed to Mealy machines that associate outputs with transitions. Transitions are driven by the outcomes of the previous play; using the three possible outcomes, I scored higher (H), my opponent scored at least as well as I did (L), and my opponent and I did not make a deal (F). These moves are called high, low, and fail. An example of one of these machines is shown in Figure 4.8.

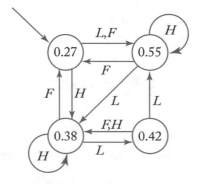

Figure 4.8: An example of a four-state automata of the sort.

The machine shown in Figure 4.8 has only four states. Because the agent has one possible payoff *per state* the number of payoffs available is relatively small. This means that agents need to have enough states to give them an acceptable number of payoffs. In addition to payoffs and transitions, the agents need an initial play. The *starting state*, shown with a sourceless arrow in Figure 4.8, yields the initial payoff. The starting state is always state zero.

The agents are represented as vectors of integers in the range 0–4,999. For an n-state agents the vector has length $4n$ with contiguous blocks of four integers specifying a state. Withing a 4-tuple (n_1, n_2, n_3, n_4) we take the first three numbers $(mod\ k)$, where k is the number of states, to obtain the transitions for play outcomes H, L, and F, respectively. The fourth number is divided by 4,999 to obtained the state label (bid) in the interval [0,1].

When updating a population, pairs of agents reproduce to replace other agents. This reproduction uses two-point crossover of the vector of integers. Mutation selects a number of loci and replaces the integers there with new ones generated uniformly at random. A parameter of the algorithm, the maximum number of mutations (MNM), controls the distribution of number of loci mutated. The number of loci mutated is chosen in the range 1-MNM, uniformly at random.

The Evolutionary Algorithm   The evolutionary algorithm is modeled on one commonly used to train game-playing agents [2, 19, 27, 31]. A population of 36 agents play a round-robin of GDD for 15 rounds. Fitness of an agent is their average payoff in those 15 rounds. The agents are sorted by score and the 24 most fit agents, breaking ties uniformly at random, are retained. The 12 least fit agents are replaced by choice pairs of distinct parents with fitness proportional selection from the agents being retained. These pairs of parents are copied over the agents being replaced and then the copies undergo crossover and mutation.

Experiments Performed   A set of five experiments were performed using the GDD defined by each of the sets given in Figure 4.7. Each experiment performed 30 independent runs of the evolutionary algorithm. The evolutionary algorithm was run for 250 generations, with a generation consisting of an updating of the population.

Summary data in the form of average and maximum score, fraction of coordinated plays, and distribution of moves made in the final distribution were saved. The final population in each generation was also saved for later analysis.

Evaluation Tools   The bids made in the final generation of each run of the evolutionary algorithm were merged to obtain an empirical distribution of bids for the evolved agents. These are plotted with scaled versions of the replicator dynamic distributions to see the degree to which adaptation by the agents moves the distribution of bids away from those generated by simple replicator dynamics. This evaluation tests the hypothesis that simple replicator dynamics are a bad predictor of the behavior of adaptive agents.

The level of coordination, for each set and run, is plotted generation-by-generation over the duration of evolution. This evaluation permits assessment of the degree to which agents are learning to coordinate their bids efficiently as well to permit assessment of the impact of the subsidies on agent coordination.

All of the sets with subsidies permit higher joint agent scores than the classical game. The best scores the agents receive, averaged over the course of evolution, give an assessment of the degree to which payouts can vary for different sets. Wasp plots for this average best fitness in each run are used to compare agent performance on the different sets.

4.2.3   RESULTS AND DISCUSSION

Figure 4.9 shows gray-scale plots of the level of coordination the agents achieved over the course of evolution for each version of GDD used in this section. Moderate levels of coordination are visible near the beginning of each set of runs. The lowest level of coordination appears in the classical Divide-the-Dollar game, suggesting that the subsidies make coordination easier. For the three simple subsidies this is exactly what one would expect because the subsidies actually increase the area of the set. The level of coordination also goes up, for the simple subsidies, as the subsidy grows larger.

The complex subsidy exhibits a larger level of coordination than the classical game, but a lower one than the simple subsidies. Since the complex subsidy has the smallest area, this suggests that the improvement in coordination is best explained by the way the subsidy encourages bids near the center of the distribution of acceptable bids. The simple subsidies perform this function in a less complex fashion fashion. The central tendency induced by a subsidy makes it easier for the bids to coordinate at a high bid value—which increases the ease of evolving high-fitness pairs of bids.

Agents learning to play the game with the complex subsidy must lean to avoid the penalized regions of low bids. Evolution apparently can learn to avoid the low regions without much difficulty, though this, together with the sets smaller area, might explain the somewhat lower level of coordination of the complex subsidy in comparison to the simple subsidies. Nevertheless, the fact that the complex subsidy has a lower total area but a higher level of coordination than set for classical Divide-the-Dollar strongly suggests that subsidies encourage coordination.

Additional evidence about the effect of the subsidies on evolving agents is given in Figure 4.10. The figure shows the distribution of bids resulting from replicator dynamics, scaled to permit comparison, with the empirical distribution of bids that arose in the final generation of evolution, merged across all 30 runs. The spiky character of the evolved bids results from the fact that the agents have only 12 bids available, each, and so repeat particular bids quite often. In evolved populations, agents will also share bid values because of common ancestry.

A number of facts can be derived from examination of Figure 4.10.

- The distributions of agent bids are not well predicted by the replicator dynamic distributions. The agents manage evolve to positions that are advantageous relative to the replicator dynamic distributions.

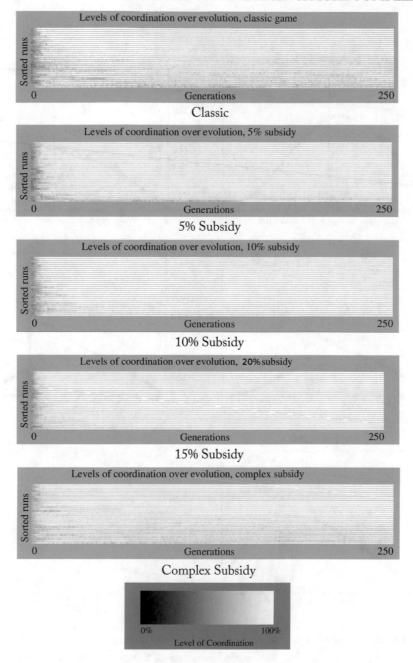

Figure 4.9: Levels of coordination for each of the 30 runs of the evolutionary algorithm performed for each set. Evolutionary time in generations is plotted left-to-right and the level of coordination is shown via gray-scale.

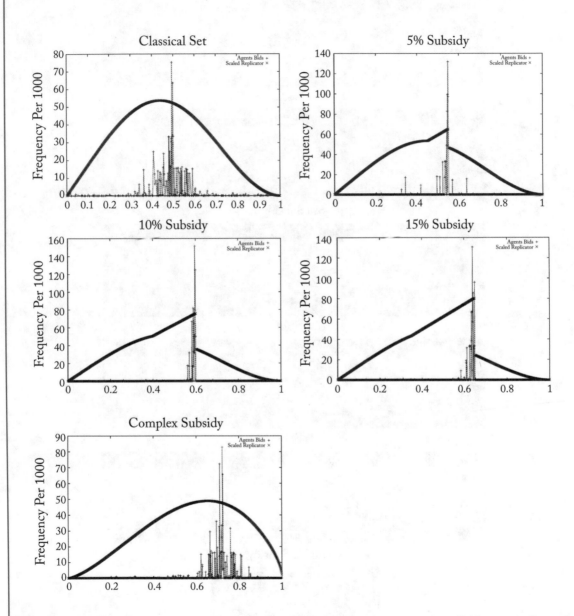

Figure 4.10: Distributions of agent bids captured in the final generation across all 30 runs performed for each set. They a plotted together with scaled versions of the distributions of bids resulting from replicator dynamics to permit comparison.

- For the three simple subsidies—the ones with discontinuities—the best bids are those just short of the discontinuity. The evolved agent bids are strongly clustered just below the discontinuity, suggesting efficient adaptation to the problem. The clustering appears to grow stronger as the size of the simple subsidy increases.

- Subsidies are successful at increasing coordination, both my making more resources available, but also, based on the behavior for the complex subsidy, be creating a favored range of bids that makes the discovery of profitable, coordinated bids easier.

Figure 4.11 shows the distribution of best fitness values, averaged over the course of evolution, for each type of GDD used in this work. Comparing this figure to the values in Table 4.3,

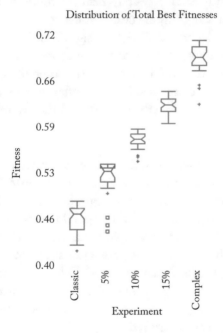

Figure 4.11: Wasp plots of the distribution of the average best fitness values for 30 runs of the evolutionary algorithm, for each of the 5 versions of GDD using in this section.

the differences in the fitness are much larger than the change in areas of the sets. For the classical and simple subsidy sets, the cost of the subsidy goes up $0.07 between the classical and the smallest simple subsidies and between adjacent pairs of simple subsidies; this amount is close to the difference between the means of the distributions. The very high cost of the complex subsidy, $0.56 does not match the increase of the scores for the complex subsidy above the classical game. This may be in part because of the penalized region of low scores.

Another deduction that follows from Figure 4.11 is that the highest scores earned by the agents are in the part of the subsidy area that yields the highest payoff. The average high payoff of about $0.47 for the classic game agrees with results from an earlier study on Divide-the-Dollar [2]. The significantly higher payoffs for all subsidies shows that the subsides have a significant impact on the types of agents that emerge from evolution.

4.2.4   CONCLUSIONS AND NEXT STEPS

In this section we introduce a generalization of Divide-the-Dollar that permits the situation modeled by the game to be controlled by the selection of the set of points that allow agents a non-zero payoff. The points in this set represent the collections of joint bids that permit the agents to make a deal. The sets used here modeled two types of subsidies. Three versions of a simple subsidy that transparently encourages bids near the center of the range were tested as well as a more complex subsidy that smoothly encourages bids nearer to the center of the distribution.

The four different subsidies tested all yielded significantly different agent behaviors. This suggests that the evolved agents were adapted to the economic environments represented by their sets. An interesting followup would be to test each type of agent on each type of set and pairs of types of agents on each of the sets. The distributions of agent bids showed that the agents adapted to their sets much more efficiently than if they were following simple replicator dynamics.

Dynamic Payoffs   Population-based techniques like evolutionary algorithms maintain populations which can find multiple optima and which tend to scatter about a given optima. This means that when we have a dynamic optimization problem [71, 90], one in which the optima change over time, and evolutionary algorithm is a good choice. The scatter of the population about optima permits them to detect and exploit the motion of the optima, so long as that motion does not jump entirely outside of the population's scatter.

In research on evolving agents to play Prisoner's Dilemma, it was found that different payoff matrices caused different types of agents to evolve [18]. The current set-based generalization of Divide-the-Dollar makes it easy to vary the game over time. The set defining payoffs for the game could be changed abruptly or morphed.

A feature of government subsidies is that they are unreliable. An election occurs and the set of available subsidies can change in an abrupt and unpredictable fashion. If we used one of the subsidy sets for part of evolution and the classic Divide-the-Dollar set for the other part, this would permit us to track the behavior of agents when a subsidy was abruptly removed or suddenly made available.

Asymmetric Games   All of the sets used in this study were chosen to be symmetric. This means that the player supplying the x coordinate of a point had no particular advantage or disadvantage relative to the player supplying the y coordinate. In this initial study of GDD, this choice avoided opening more than one can of worms. There is, however, no reason that asymmetric sets cannot be used.

The set shown on the left side of Figure 4.12 has an asymmetry that models a structural inequality strongly favoring the player supplying the x-coordinate. This example shows that GDD can be used to model structural or legislative inequalities and emphasizes the design simplicity resulting from specifying the payoffs of the game with a set.

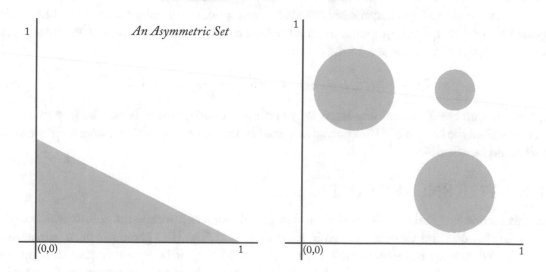

Figure 4.12: Shown on the left is an asymmetric set that strongly favors the player supplying the x-coordinate. Shown on the right is a disconnected scoring set the simulates having different sectors for a company to work in.

An additional issue arises when an asymmetric set is used; it matters a great deal which player is supplying what coordinate of the set. One way of dealing with this is to use multiple populations as in [27], although the goal there was not dealing with asymmetry. The *multiple worlds model* [42] was designed for game-theoretic simulations in which there are multiple actors with asymmetric roles. This model would be a natural choice for managing multiple populations of agents in an asymmetric GDD.

Disconnected Sets   In a complex regulatory environment, or one with arbitrary regulatory costs, the fact that fees are triggered by particular expenditure levels makes it possible to have a payoff scheme in which the sets of coordinated bids need not be connected. The set shown on the right side of Figure 4.12 is an example of disconnected set of this type. Disconnected sets create separate collections of effective pairs of behaviors. This means that the evolution of agents on sets of this type would be more complex than that for simple connected sets.

More Than Two Players   All the sets used to specify payoff schemes for GDD games in this study are two-dimensional and so specify two-player games. An set in n-dimensions specifies and n-player game. The set

$$D = \{(x_1, x_2, \ldots, x_n) : x_1 + x_2 + \cdots + x_n \leq 1\}$$

with all coordinates positive, would be a very simple generalization of Divide-the-Dollar to n players. If the sum of the player's bids totals at most a dollar, the players get their bids. This game has the perhaps annoying feature that the bids must, on average, have size $\frac{1}{n}$. This makes a natural choice to examine sets of the form:

$$D_S = \{(x_1, x_2, \ldots, x_n) : x_1 + x_2 + \cdots + x_n \leq 1 + S\},$$

where the number S is a subsidy intended to promote multi-partner deals. The impact of the values of S on the level of coordination among the agents is an example of an interesting question that could be asked.

4.3 THE SNOWDRIFT GAME

In this section we study the *Iterated Snowdrift* (ISD) game and review the results which Garrison Greenwood and Chopra showed in their work [62]. The IPD game has been extensively investigated to help researchers gain a better understanding of how cooperation develops in populations. One criticism of IPD is it underestimates the level of cooperation particularly in human populations. The ISD game has emerged as a viable alternative model, in part because it predicts higher cooperation levels. To date, no numerical analysis of ISD has been done. In this section we study the results from a numerical analysis conducted on an ISD with an N-player, well-mixed population.

Cooperation is ubiquitous throughout nature, but the underlying mechanisms are not well understood. Defection often pays higher dividends when others cooperate but no one derives any benefit if everyone defects. This conflict is referred to as a social dilemma [51]. Game theory offers a framework to study cooperation and the conditions under which it develops. One of the most widely studied game in this area is the IPD. In IPD, two players choose whether to cooperate (C) or defect (D). If one player plays C and the other plays D, the cooperator incurs a sucker cost S while the defector gets a temptation payoff T. Mutual cooperation results in both players getting a reward payoff C. Mutual defection gives both players a punishment payoff D. The two payoff constraints are $T > C > D > S$ and $2C > T + S$. In a single encounter defection is always the best strategy. However, some studies have shown that in IPD, where repeated encounters occur and players can respond to past player behaviors, it is possible to achieve some levels of cooperation under certain circumstances [39, 54].

The Snowdrift (SD) game, also known as Hawk-Dove or Chicken game, has emerged as an interesting alternative game for studying cooperative behavior. In the one-shot SD game the best strategy is defection if the other player cooperates, but worse if the other player defects [82]. The SD game differs from PD by a reordering of the payoffs, i.e., $T > R > S > P$.

The ISD game, where individuals interact repeatedly, has attracted a growing interest because it allows cooperation to persist, which is quite common in nature. Indeed, one of the complaints with IPD is it predicts much lower levels of cooperation in humans than what is typically found. On the other hand, ISD predicts higher human cooperation levels [57].

Numerous experiments have been conducted on ISD in recent years. Some work has been done with well-mixed populations [106], but the majority has been done on spatial graphs where only limited interactions are allowed. The rationale behind this is in many biological systems individuals can only interact with other individuals who are within some small Euclidean distance of each other. ISD work with spatial graphs includes two-dimensional lattices [65], random networks [89], small-world networks [58], and scale-free networks [103]. In this work, however, we focus on well-mixed populations.

Unfortunately, little insight resulted from much of this previous work because the authors provided only weak explanations of their results. Essentially, all these previous ISD researchers did was define the interaction limits, define the terms of how evolution occurs, and then report the long-term results of cooperation in the population. For example, in [65] each player was placed on a unique node of a two-dimensional square lattice with periodic boundary conditions. Players can only interact with their N, S, E, and W neighbors. All players were initially assigned a value of "0" or "1" with equal probability. During each round a player x randomly chooses a neighbor y. Player x adopts the strategy of player y with a probability $f(P_y - P_x)$ where P_x (P_y) is the cumulative payoff of player x (y) and $f(z) = 0$ if $z < 0$ and $\frac{z}{T-S}$ otherwise. The outcome of these encounters are recorded and all updates are done simultaneously. Their results indicate IPD tends to form large clusters of cooperators whereas cooperators in ISD form a dendritic skeleton.

Our objective was to go beyond these previous studies by doing a qualitative analysis of ISD dynamics. We investigated an N-player ISD game with a well-mixed population and pairwise player interactions. Our results indicate the game dynamics are remarkably sensitive to the cost of cooperation. In fact, for particular cost values nearly half of the population become cooperators and this proportion persists even though player strategies continue to evolve.

Surprisingly, an extensive literature search failed to turn up any numerical analysis of ISD. Our intent was to conduct such an analysis but the obvious question is where to start. Fortunately, an extensive numerical analysis has been conducted on a related game [44]. The minority game (MG), although quite different from ISD, has some fundamental similarities, which means we can pattern our numerical analysis after that which was done on the MG.

4.3.1 THE GAME MODELS

Since the numerical analysis conducted on MG is the basis for our analysis of ISD, it is important to describe both games, which is done in this section. Both games have pairwise interactions among players in a well-mixed population. In the original description of MG players chose "0" or "1" but this could easily be changed to picking C or D. Let C and D be represented by the two-dimensional unit vectors $s = \begin{pmatrix} 1 \\ 0 \end{pmatrix}$ and $\begin{pmatrix} 0 \\ 1 \end{pmatrix}$, respectively. Then at the end of each round a player x in a population ω of size N receives a payoff of $payoff(x) = \sum_{y \in \Omega, y \neq x} s_x^T Q s_y$ where Q is the payoff matrix. The total payoff to x is the sum of the payoffs accumulated over all rounds of play. Both MG and ISD compute payoffs this way.

Readers should pay particular attention to the similarities between the evolutionary MG and the evolutionary ISD which are discussed in this section.

The Minority Game   The minority game was originally introduced by Challet and Zhang [43] to help study adaptive behavior in a population of interacting players. MG is a well-mixed game where during each round every player plays against the other $N - 1$ players ($N \geq 3$). At the beginning of each round all players simultaneously choose either "0" or "1". There will always be a minority choice since N is always an odd number. The payoffs are simple: those in the minority get a payoff $b > 0$ while those in the majority get a payoff of zero. Payoffs are awarded at the end of each round. Players have only limited information about the previous M rounds of play. They do know what the minority choice was, but they aren't told the size of that minority. Players also don't know what choices other individual players made. Thus, the MG provides no framework for studying behavior based on altruism or reciprocity like some other games do [48]. The basic MG model gives each player s strategies to choose from.[1] Rather than randomly choosing a strategy to use, at each time step a player picks the strategy that theoretically would yield the highest cumulative payoff since the beginning of the game. This pick is made in a very straightforward way. A strategy is awarded a "virtual point" if playing that strategy would have put the player in the minority. Strategies can get virtual points even if they were not played. At each time step the player chooses the strategy with the highest number of virtual points. A player receives a "real point" if the strategy picked for play puts the player in the minority in the current round. The real point total serves as a fitness measure; highly fit players are frequently in the minority because they have good strategies.

Table 4.4 shows one possible pair of strategies for a memory size of $M = 3$. The two choices are denoted by "0" and "1" and the table indicates what the next choice (or play) should be for a given history of M outcomes. For example, if the last three minority outcomes were "1 0 1," and the player chooses strategy s0, then at the next time step this player would play "0". However, if strategy s_1 is picked then this player would play "1" (for the same history). With a memory size of M there are 2^M possible previous outcomes. This means there are 2^{2^M} possible strategies.[2]

Some interesting and quite unexpected game dynamics occur when strategies can evolve during the game. Challet and Zhang [50] later introduced the so-called evolutionary minority game where players may modify their strategies. Each player has $s = 2$ strategies. Both strategies are evaluated at each time step but only the best performing one is actually used. Players can

---

[1]Previous studies indicate $s = 2$ is sufficient to capture the dynamics of the game.
[2]A search space for $M = 5$ has over 4 billion strategies.

Table 4.4: One possible set of strategy pairs with memory size $M = 3$

Previous 3 Outcomes	Next S_0	Play S_1
000	0	1
001	1	0
010	1	1
011	0	0
100	1	1
101	0	1
110	0	0
111	1	0

adapt by changing to a different strategy every few time steps. At periodic intervals the poorest performing player replaces his two strategies by cloning the two strategies of the best performing player. The new strategies are then mutated with some small probability. The virtual points totals for both cloned strategies are reset to zero so learning can take place. Figure 4.13 shows the long-term outcome from a typical MG run. Here "Ones" represents the number of players who chose "1" in each round for a population of size $N = 1001$. Notice initially there is a large variation but eventually the population settles down to small variations around $N/2$. This tapering indicates a learning process has taken place. We found nearly identical dynamics can arise in the Snowdrift game under certain cost-to-benefit ratios. This aspect will be discussed in detail in Section 4.3.

A thorough literature search found subsequent work on the evolutionary MG model contributed little beyond Challet and Zhang's original work. Several researchers published papers where the only thing they did new was to replace cloning with a genetic operator. For example, Sysi-Aho et al. [83] used 1-pt crossover on the s strategies to get the replacement strategies while Yang et al. [104] used a cut-and-splice operator. Neither paper offered any new insights into minority game behavior.

The Snowdrift Game   Picture two drivers who are stuck on opposite sides of a snowdrift. The only way they can get free is to shovel the snow to remove the snowdrift. A driver cooperates by agreeing to get out of his car and shovel snow. Conversely, a driver defects by deciding to stay inside his car where it is warm and dry instead of shoveling snow. Cooperation pays a benefit $b > 0$ (freeing the car) but the effort of shoveling snow has a cost $c < b$. If both drivers cooperate this cost is shared equally. The best option is to let the cooperator do all the work because the defector still gets the benefit b but the cooperator pays the full cost c. Of course if both drivers defect then they remain stuck in the snowdrift and neither one gets any benefit. The SD payoff matrix is as shown in Table 4.5. In the iterated version players interact over a sequence of rounds.

Table 4.5: SD payoff matrix

	C	D
pay off to C	$b - \frac{c}{2}$	$b - c$
pay off to D	b	0 .

A restriction applies here as it does for IPD: there is a finite number of rounds of play, but players are not told that number beforehand. In this work we are only interested in the Iterated SD (ISD) version of the game.

For the N-player version each player plays against the other $N - 1$ players with only the focal player getting the payoff shown in Table 4.5. (The opposing player receives nothing at that time but must wait until he is the focal player.) Players receive nothing in the all-defect case. With $k > 0$ cooperators, all cooperators receive a payoff of

$$P_C(k) = (b - c/2)(k - 1) + (b - c)(N - k) . \tag{4.5}$$

The first term is the payoff for playing against the $k - 1$ other cooperators while the second term is the payoff for playing against the $N - k$ defectors in the population. All defectors receive a payoff of

$$P_D(k) = b\dot{k} . \tag{4.6}$$

These equations indicate a small number of defectors can prosper, which might tempt others to defect. Of course if every player decides to defect then the entire population suffers. N-player ISD therefore qualifies as a social dilemma [51].

Similarities Between MG and ISD   Both MG and the N-player ISD have an odd number of players (with $N > 3$), so there is always a minority choice in each round. The MG players who made the minority choice are the only ones who receive a positive payoff in a given round. Consequently, it does not matter if the minority choice is C or D, the minority always gets a higher payoff. Intuition says this same outcome should also occur in the N-player ISD game. Suppose the minority choice is C. According to payoff matrix 4.5, each cooperating player receives a positive payoff from all $N - 1$ other players while the defecting players receive a zero payoff from most of the interactions. Conversely, if the minority choice is D, the defecting players receive a high positive payoff from the majority of the population while the cooperators receive a lower payoff from most players. In either case a small number of minority players can successfully invade the population.

A simple example will help fix ideas. Suppose there are 3 players with $b = 1$ and $c = 0.2$ and payoffs are awarded using (4.5) and (4.6) as appropriate. If the players pick CCD, then both cooperating players each get $P_C = 1.7$ while the single defector gets $P_D = 2$. Conversely, if the players pick DDC, the cooperating player gets $P_C = 1.6$ and the two defectors each get $P_D = 1$. In both cases the minority player did better. However, as we will soon see, this outcome holds if and only if the cooperation cost-to-benefit ratio is low enough. MG and the N-player ISD game (with a low c value) both favor minority players. This similarity has important implications: the numerical analysis techniques used previously in [44] for MG can serve as a basis for our numerical analysis of the N-player ISD.

4.3.2   EXPERIMENTAL RESULTS

Memory Size and Variance   This set of experiments was intended to measure the cooperation variance as a function of memory size M. A population of $N = 501$ players, each with $s = 2$ strategies, were randomly initialed for various values of M. The number of cooperating players was recorded over a $T = 10,000$ round game, which forms a time series A(t). The mean of this series is

$$\langle A(t) \rangle = \frac{1}{T} \sum_{t=1}^{T} A(t)$$

and the variance is

$$\sigma^2 = \frac{1}{T \cdot N} \sum_{t=1}^{T} (A(t) - \langle A(t) \rangle)^2.$$

Sixteen independent ISD runs were conducted with $2 \leq M \leq 10$. In each run we set $b = 1$ and $c = 2/3$. No mutation was used during this portion of the investigation. More specifically, players still choose strategies based on virtual point totals and cloning was still used to replace strategies, but cloned strategies were not mutated.

Figure 4.13 shows the results. Notice that the variance reaches a minimum M = 6. For larger M values the variance begins to increase, asymptotically approaching the random choice strategy, which has a variance $\sigma^2 = 0.25$.

It is interesting to note a similar MG investigation was conducted by Savit et al. [92]. They also did not mutate cloned strategies. Our variance results are nearly identical to their MG variance results (see Figure 4.14 in [92]).

We next determined the role mutation plays. The above experiments were repeated, but this time cloned strategies were mutated. The results are shown in Figure 4.15. Notice the variance exceeds that of the random choice strategy for large M values.

Cost-to-Benefit Ratio   These experiments were conducted to see how the cost-to-benefit ratio $(r = c/b)$ might affect the number of cooperators over time. A population of $N = 1,001$ plays an evolutionary ISD with a fixed $b = 1$ but with different c values. (This normalization makes $r = c$ and allows us to focus on the cost-to-benefit ratio.) All runs were conducted with memory size $M = 5$. Figure 4.16 shows the number of cooperators $A(t)$ over 10,000 rounds with $r = 3/4$. The number of cooperators initially decreases toward $N/2$ but soon thereafter the population enters a phase where defectors gradually take over the entire population. In sharp contrast is the situation when $r = 2/3$, which is shown in Figure 4.17. Now the number of cooperators drops to a level with a small variation around the mean $0N/2$ and remains there. This is a quasi-stable

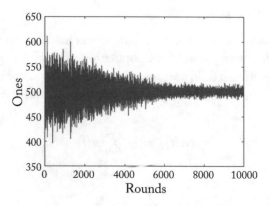

Figure 4.13: Temporal attendance for the genetic approach with $N = 1,001$. The tapering indicates a learning process.

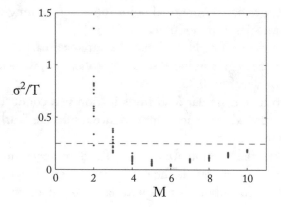

Figure 4.14: Variance as a function of memory size M in a time series of ISD cooperators over $T = 10,000$ rounds. In the random case, i.e., where players choose C or D independently and with equal probability the time series is formed from binomial sampling and the normalized variance becomes $\sigma^2/T = 1/4$, which is indicated by the dashed line. Strategies were not mutated.

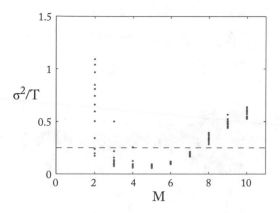

Figure 4.15: Variance as a function of memory size M in a time series of ISD cooperators over $T = 10,000$ rounds. The dashed line indicates the variance for the random choice strategy where players pick C or D with equal probability. Cloned strategies are mutated here as opposed to the Figure 4.14 results where strategies were not mutated.

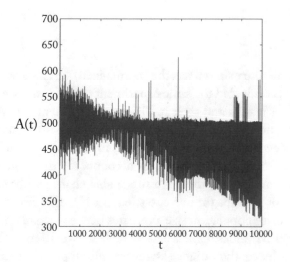

Figure 4.16: Time series of cooperators in the evolutionary ISD with $r = 3/4$.

state where roughly half of the population consists of cooperators. The reader should compare Figures 4.17 and 4.13; the evolutionary ISD dynamics with $r = 2/3$ and the MG dynamics are virtually identical.

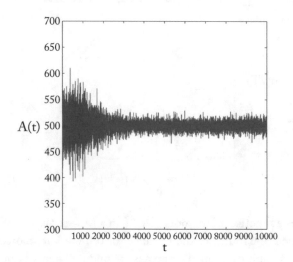

Figure 4.17: Time series of cooperators in the evolutionary ISD with $r = 2/3$.

4.3.3   DISCUSSION

Figure 4.14 shows how memory size effects the (normalized) variance of a time series of cooperators. Manuca et al. [91] conducted the same experiment on MG that we conducted to generate the Figure 4.14 plot. Our results and their results are nearly identical. We presume the strong similarity between MG and the N-player ISD (without mutation) is the underlying reason.

Manuca et al. believed below $M = 6$ there was little predictive information contained in the histories that players might exploit and this accounts for the large variation spread. One reason is with small M, say $M = 2$, there is a reasonable chance a player may have two nearly identical strategies, except perhaps for one entry (e.g., "01"). This situation could lead to cyclic behavior where the minority population size alternates between small and large deviations from the mean $N/2$. For larger M values there is considerable information in the histories but players are unable to exploit it. Hence, the variance asymptotically approaches the random choice strategy. We see no reason to believe why those same explanations won't hold for the N-player ISD (without mutation) as well. The results depicted in Figure 4.14 were obtained from non-mutated cloned strategies. Our results indicate mutation does have a dramatic effect on the variance in the evolutionary ISD. Notice in Figure 4.15, where the cloned strategies were mutated, larger memory sizes produce variance levels higher than the random choice strategy. Challet and Zhang also observed this same phenomenon in the evolutionary MG (see Figure 4.14 and accompanying text in [44]).

The long-term dynamics in the evolutionary ISD depend heavily on r. Figures 4.16 and 4.17 show radically different dynamics with the only difference being $r = 3/4$ in one case and $r = 2/3$ in the other case. Both figures show initially a large variation in the number of co-operators but the population tends to slowly self-organize around $N/2$ with a smaller variation. This self-organization, which suggests a learning process is taking place, is a natural consequence of selection pressure. The minority player in an evolutionary ISD has higher fitness. When the population contains a large number of cooperators (defectors), the defectors (cooperators) are favored for reproduction. Over time this process drives the cooperator frequency in the population toward a limit of $N/2$.

Figures 4.16 and 4.17 also show the cooperator frequency in the population may or may not stay near N/2 for long, depending on r. To help understand why, we plotted (4.5) and (4.6) for various r values. As shown in Figure 4.18, with $r = 0.2$ cooperators have a considerably higher payoff unless the population nears saturation, after which defectors have a higher payoff. Conversely, for $r = 0.7$ cooperators have an advantage only when they make up less than half of the population. When $r = 2/3$, where the interesting dynamics occur, the number of coop-erators is roughly half of the population. (This is a special case, which we will discuss in more detail shortly.)

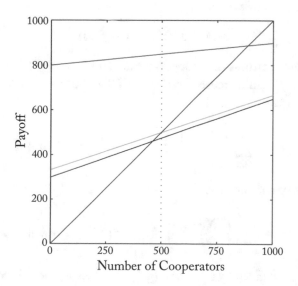

Figure 4.18: Payoffs for various cost-to-benefit ratios. The figure shows the payoff for $r = 0.2$ (blue) $r = 2/3$ (green) and $r = 0.7$ (black). The red line shows the payoff for defectors.

At first these results may appear counterintuitive. From Table 4.5 a defector always gets a higher payoff when playing against a cooperator. How then can a large number of cooperators prosper in a population dominated by defectors? There is actually a simple explanation for this: it all depends on the cost c that is paid for cooperating. The reason is with low costs cooperators receive substantial rewards regardless of whether they play other cooperators or defectors.[3] For example, with $c = 0.2$, the payoff to a cooperator playing a defector is 0.8 whereas a defector gets a payoff of 1.0 for playing a cooperator. All players receive $N - 1$ payoffs per round because they interact with $N - 1$ other players. But remember a defector gets nothing for interacting with another defector while a cooperator gets something from everyone he plays against. When the payoffs of (C, C) are nearly as high as the payoffs for (D, C), a cooperator can acquire large enough payoffs to be considered highly fit and therefore survive—even if most players he encounters are defectors.

Figure 4.16 shows the population dynamics when the cost-to-benefit ratio is high ($r = 3/4$). As the population self-organizes, driving the number of cooperators toward $N/2$, the fitness increases for defectors. This increase quickly turns the cloning process into a positive feedback force, which disrupts the self-organization by reproducing even more defectors in the population. We observed just the opposite effect, i.e., rapid growth in cooperators when r is small. This cooperator growth is expected because the cost to cooperate is low.

But that still doesn't explain what is so special about the case when $r = 2/3$. To help gain some insight we can use replicator dynamics. Let k be the number of cooperators in a population of size N. Then $z(t) = k/N$ is the frequency of cooperators at round t. Following the derivation in [106], the time evolution of $z(t)$ is

$$\dot{z}(t) = z(t)(f_C(t) - \overline{f}(t)), \tag{4.7}$$

where $f_C(t)$ is the average fitness of a cooperator at round t and $\overline{f}(t)$ is the average fitness of the entire population. $\dot{z}(t)$ will increase if $f_C(t) > \overline{f}(t)$ and decrease otherwise. With binomial sampling (and dropping the t for clarity)

$$f_C = \sum_{k=0}^{N-1} \binom{N-1}{k} z^k (1-z)^{N-1-k} P_c(k+1) \tag{4.8}$$

and the average fitness of a defector is

$$f_D = \sum_{k=0}^{N-1} \binom{N-1}{k} z^k (1-z)^{N-1-k} P_D(k+1), \tag{4.9}$$

where $P_C(\cdot)$ and $P_D(\cdot)$ are given by (4.5) and (4.6). The average population fitness is

$$\overline{f} = z f_D + (1-z) f_D . \tag{4.10}$$

---

[3]On the other hand, defectors get the same payoff ($b = 1$) for playing a cooperator regardless of the cost c.

Substituting (4.10) into (4.7) yields

$$\dot{z}(t) = z(1 - z)(f_C - f_D) . \tag{4.11}$$

This system evolves to a steady-state, i.e., a Nash equilibrium when $\dot{z} = 0$, which occurs when $f_C = f_D$.

Consider now the special case when $r = 2/3$. The population tends to naturally self-organize around a mean of $k = N/2$ cooperators (see Figure 4.17). As $k \to N/2$, the payoffs for cooperation and defection become the same. Thus, $\dot{z} \to 0$ as the population self-organizes. The population continues to evolve, but selective pressure tends to keep the population in this quasi-stable state. Any variation above (below) $N/2$ cooperators gives a higher payoff to defectors (cooperators) driving the population back to $N/2$ cooperators.

Of course when $r = 2/3$ then the critical number of cooperators must change to put the population into a steady state. But with $r = 2/3$ this critical number is $N/2$, which just happens to also be the natural self-organization target selection pressure tries to drive the population toward. We found this sustained self-organization occurs independent of the initial distribution of player strategies.

This critical number is the only number where the quasi-stable behavior occurs because it is the only value that makes $\dot{z}(t) \approx 0$. Figure 4.19 shows the long-term evolutionary ISD behavior when $r = 2.01/3$. Notice the population still self-organizes, but eventually the variance starts to increase. It is easy to show $|\dot{z}(t)|$ is slightly larger when the cost is $2.01/3$, which weakens the quasi-stable state.

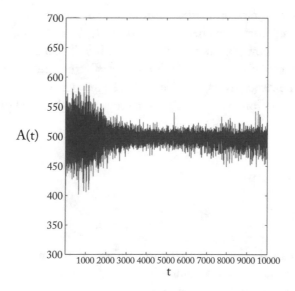

Figure 4.19: Time series of cooperators in the evolutionary ISD with $r = 2.01/3$.

4.3.4   CONCLUSIONS

Our experiments and qualitative analysis on the N-player evolutionary ISD resulted in several noteworthy conclusions.

- *With payoffs* (4.5) *and* (4.6) *a small number of cooperators will always have a higher payoff than a large number of defectors. This is a phenomenon associated with the N-player ISD; a defector always does better than a cooperator in the 2-player ISD.*

- *The evolutionary ISD has a critical value of cost-to-benefit ratio where the average cost of co-operation equals the average cost of defection. At this value the cooperator frequency reaches a quasi–stable state.*

- *A large number of cooperators can have an advantage over a smaller number of defectors if the cost is low enough.*

- *With* $r = 2/3$ *and payoffs defined by* (4.5) *and* (4.6)*, the evolutionary ISD enters a prolonged quasi-stable state independent of the initial distribution of strategies. Replicator dynamics pro-vides a plausible explanation.*

- *The variance in a time series of cooperators reaches a minimal value with a memory size* $M = 6$.

- *For* $3 \leq M \leq 7$ *this variance is lower than the variance of the random choice strategy. Above* $M = 7$ *the variance is higher indicating large memory sizes are disruptive.*

In our future work we plan study the information content in the histories to gain a better understanding of how memory size affects behavior. Larger memory should contain more in-formation, but is not clear how players might exploit it. It will also be interesting to see how the evolutionary ISD dynamics change with different payoff structures. For example, in [106] all cooperators get the same fixed payoff, but the cost is shared amongst all of them. That is, with k cooperators in the population, each (C, C) interaction gives the focal player a payoff of

$$P_C = b + \frac{c}{k}$$

and the (D, C) interaction pays

$$R(x) = \begin{cases} 0 & k = 0 \\ b & k > 0. \end{cases}$$

The evolutionary ISD can also be generalized to a public goods game. Here cooperators would contribute to a public good, which is commonly shared, but each additional contribution has lower worth. The cost-to-benefit ratio determines whether defectors or cooperators are dom-inant [82]. In our future work we intend to conduct a numerical analysis of the public goods game version of ISD because of its relevance to climate change programs.

CHAPTER 5

Noise!

This chapter introduces the noisy IPD and uses finite state machines, one of the more versatile representations from the work presented in Chapter 2, to investigate the effect on cooperative behavior of the presence of noise. Noise make the game the agents are playing more complex and difficult—but as we will see it also has a huge impact on the agent's behavior. Noise models misunderstanding, something that is common in human interaction, and so is a natural and necessary subject when trying to understand training game-playing agents with evolution.

As part of the analysis or the behavior of agents with and without noise, a new quantity, evolutionary velocity, is introduced and is used to expose an unexpected and critical difference between agents evolved with and without noise using the finite state representation. Computation of evolutionary velocity uses the fingerprints introduced in Chapter 2.

5.1 NOISY GAMES ARE DIFFERENT

The type of noise used in this chapter creates a disjunction between perceived actions and payoffs. We introduce a small chance that an agents misunderstand the other player's last move. If two agents both cooperate during a noise event, then the agent affected by the noise will react as if their opponent had defected, but will nevertheless receive a payoff for mutual cooperations. If the agent based their actions on the payoff then this would act to circumvent the noise; this also suggests a second type of noise, one that actually changes the action and not just the perception of the action.

In [25] it was found that evolving agents to play the Iterated Prisoner's Dilemma for a long time gave them a substantial competitive advantage against agents evolved for less time from different evolutionary lines. This phenomenon, called *non-localized adaptation*, suggests that agents are gaining not just skill playing the agents with whom they are co-evolving but general skill at playing the Prisoner's Dilemma; they have a statistically significant advantage against agents they have never been evaluated against before.

In [26] it was demonstrated that non-localized adaptation takes place in a steady fashion across much of evolution. This work presented here extends measurement of competitive advantage to include the effects of noise as well. Non-localized adaptation is also observed in other co-evolutionary contexts. In [56], it was observed in competitive exclusion in a spatial model of plant growth. In [9], it was observed in populations of virtual robots evolving to paint a floor in two competing colors. The effect was observed in predator-prey models in [6] and in populations of agents in Divide-the-Dollar in [10].

Fingerprints were defined in Chapter 2. This chapter uses an approximation to the fingerprint function consisting of 25 numbers which are the values of the fingerprint function on a 5 by 5 evenly spaced grid from 1/6–5/6 is used. Given that fingerprints are functions they are a subset of a potentially infinitely dimensioned space, making the use of a 25-point sampling grid potentially suspect. In [36], it is proved (Corollary 2) that the space of fingerprint functions is at most six-dimensional, making a 25-point sample a generous one.

One application of fingerprints is to place a metric-space structure on the space of Prisoner's Dilemma strategies. The Euclidian distance between the 25-sample approximations is used in this chapter. This distance forms the basis of estimating the evolutionary velocity of populations.

5.1.1   EXPERIMENTAL DESIGN

The agent representation used here is 8-state finite state machines with actions associated with transitions between states (Mealy machines), as described in Chapter 2. The crossover and mutation operators used also follow those from Chapter 2. The evolutionary algorithm used operates on a population of 36 agents, a number chosen for compatibility with previous chapters. Agent fitness is assessed by a round-robin tournament in which each pair of players engage in 150 rounds of the IPD.

Three collections of runs were performed with different versions of this round-robin tournament. The first, serving as a baseline, had perfect communication between the agents. The second and third had an average of 1% and 5%, respectively, of actions misunderstood by the opponent. In the implementation used here, the agent is represented as a finite state machine that bases their response solely on the opponent's perceived action, and so the disjunction of apparent action and payoff does not interdict the effect of the noise.

Reproduction is elitist with an elite of the 24 highest scoring strategies, another choice that maintains consistency with past studies. When constructing the elite, ties are broken uniformly at random. Twelve pairs of parents are picked by fitness-proportional selection with replacement on the elite. Parents are copied, and the copies are subjected to crossover and mutation.

In each simulation the evolutionary algorithm was run for 6,400 generations with 100 replicates (experiments with distinct random number seeds). The elite portion of the population in generations 50, 100, 200, 400, 800, 1,600, 3,200, and 6,400 was saved for analysis. These sample points are termed *epochs*. This yields 100 sets of 24 machines at each of eight epochs for each of the three noise levels. A number of descriptive statistics are saved for each generation of each replicate. These include the mean fitness, a 95% confidence interval on the fitness, the maximum fitness, and the evolutionary velocity of the population, defined in Section 5.2.

5.1.2   ANALYSIS TECHNIQUES

Probability of Cooperativeness   For each epoch and noise level, the fitness (average Prisoner's Dilemma score) of the saved elite population in a noise-free tournament was computed. Following [14, 21] a population was judged to be *cooperative* if its average score was at least 2.8 out of a maximum possible of 3.0. The fraction of populations that were cooperative was modeled as the parameter of a binomial distribution; Figure 5.3 summarizes the results. The choice to evaluate all populations without noise permits an apples-to-apples comparison of the populations evolved at different noise levels. This is the only instance in this work where fitness was recomputed without noise.

Competitiveness and Non-Localized-Adaptation   While both competitiveness and non-localized adaptation were analyzed in [31] a different form of these analysis are performed here. The results are in substantial agreement, but the method of analysis has been revised and the presentation of the information has been substantially modified in a manner that enhances clarity.

Competitiveness is measured between sets of agents evolved using different fitness evaluation processes. It is intended to assess the impact on the agent's competitive ability of those distinct fitness evaluation processes. Three different fitness evaluation processes are used in this section, all based on round robin tournaments, but using different levels of noise. We evaluate competitiveness in the zero-noise environment—a practice that would logically give the agents evolved in that environment an advantage. The zero noise environment was chosen because it yields deterministic results on any given pair of agents. As we shall see, agents evolved without noise did not benefit from its lack when evaluated against agents evolved in the presence of noise. Non-localized adaptation is measured between agents from distinct epochs but evolved using the same fitness evaluation process. It measures the impact of additional evolution.

Save for the identity of the sets of agents being compared both competitiveness and non-localized adaptation are measured in the same fashion. All agents for a given noise level and epoch are loaded into agent-pools. All pairs of distinct agent-pools are then compared. The comparison is performed by sampling 10,000 pairs of agents, each pair consisting of one agent from each of the agent-pools being compared. Each pair plays 150 rounds of IPD. Each time an agent from the first pool outscores an agent from the second pool a *success* is recorded. Each time an agent from the second pool outscores an agent from the first a *failure* is recorded. The successes and failures are modeled as Bernoulli experiments using a normal approximation to the binomial distribution. The Bernoulli parameter being estimated is the probability an agent from the first pool will beat an agent from the second.

A 95% confidence interval is constructed for the Bernoulli parameter for each pair of agent-pools. Examples of the presentation of these confidence intervals for competitiveness appear in Figure 5.4 while those for non-localized adaptation appear in Figure 5.6.

Two Forms of Competitiveness   Non-localized adaptation and competitiveness both measure the ability of two agent types to outscore one another in a pairwise competition. Axelrod's famous tournaments [38, 39] measured a completely different sort of competitive ability. This is the ability to have the highest score in a tournament with a diverse collection of other players. The strategy TFT cannot beat another player in a pairwise contest. The only way to get ahead of another player is to defect against them when they cooperate. TFT never defects first and so always defects against its opponent's cooperation on a move directly after the opponent does the same. In spite of this, TFT won both of Axelrod's tournaments. This is because, even though it was outscored by all its opponents, it elicited cooperation in many of them. Its opponents, on the other hand, sometimes got into very low-scoring sequences of defections with one another. At the end of the tournaments TFT had the highest total score.

We create a contest similar to a contributed agent contest in the following fashion. A contest consists of a round-robin tournament with 36 agents, the number used in fitness evaluation during evolution, using 150 rounds of IPD. In order to prevent genetically based collaboration, no two agents in a given tournament were selected from the same replicate. Twelve agents were selected from each of the three available noise levels and all agents in a given tournament were selected from the same epoch. The agents were selected from the saved elite population uniformly at random and the replicates used in a tournament were also selected uniformly at random, without replacement. A set of 10,000 tournaments were performed. For each epoch 95% confidence intervals for the mean population rank of the agents from each noise level were computed.

Fingerprint Analysis: Voronoi Tiles   In order to analyze Prisoner's Dilemma strategies one can try to find some way to define and name each individual strategy. Table 3.1 lists 12 strategies with names. However, naming quickly becomes intractable as the number of strategies explodes. Naming also gives no objective way to compare the strategies. A Prisoner's Dilemma fingerprint can be used to compare strategies by determining how far apart they are. Graphs of the full fingerprint functions for three strategies are shown in Figure 5.1. These are functions from the unit square to the range from 0–5 (from S–T).

In order to analyze the evolved strategies, 12 reference strategies, given in Table 3.1, were selected. These strategies represent all unique fingerprints that appear in the collection of look-up tables for Prisoner's Dilemma based on the opponent's last two moves [36]. They are all equivalent to one- or two-state finite state machines. A stylized depiction of their relative positions in fingerprint space is shown in Figure 5.2.

The evolved strategies trained in this work are analyzed in terms of which of these strategies they are closest to. This technique is called a *Voronoi tiling* or a *Dirichlet tessellation*. Explanations of the technique can be found in many places; one good one is [53]. If a strategy is closer to Strategy S than it is to any other reference strategy, then it is said to be in the Strategy S tile. The assumption is that strategies with fingerprints that are close together behave similarly, so we can group them together and analyze them in terms of the simple strategies that we understand well and in terms of where they fall in fingerprint space.

The space of fingerprint functions might be as much as six-dimensional. However, the error is small when strategies are projected onto two dimensions using non-linear projection, a technique defined in [17] and applied in [32] to RNA folding. The technique uses an evolution-

Figure 5.1: Fingerprints for tit-for-tat (left), tit-for-two-tats (middle), and Fortress 3 (right). Black indicates 0(S), white a value of 5(T) except for three additional light-colored areas. The lower left shows when scores are close to 1 (D), the middle shows when scores are close to 2.25 (random play) and the upper right shows when scores are close to 3 (C).

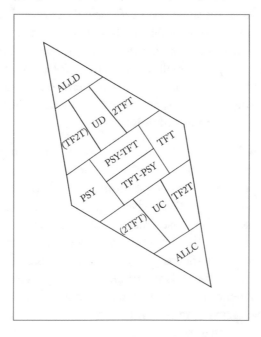

Figure 5.2: Stylized depiction of tiles, labled with their center strategies, in fingerprint space.

ary algorithm to project multi-dimensional points onto a two-dimensional picture preserving the distance relationships between the points as closely as possible. Thus, it is reasonable to conceptualize the space in two dimensions. The greatest possible fingerprint distance between any two strategies is that between ALLD and ALLC (proved in [36, Theorem 3]); the next greatest orthogonal distance is between TFT and Psycho. One can visualize fingerprint space as the interior of the diamond formed by these four strategies, as shown in Figure 5.2.

5.1.3 RESULTS AND DISCUSSION

In an early study of the impact of noise on the IPD [85] it was found that higher levels of noise yielded lower levels of cooperation. This study evolved 16-state finite state machines using a binary version of the Moore encoding for 50 generations with the same noise levels used in this study. The top panel of Figure 5.3 demonstrates that allowing additional evolutionary time changes this result. In the final epoch samples, and only in that epoch, the significant differences between the probabilities that a population will be cooperative at different noise levels disappears. The noise free and 1% noise populations never exhibit a significant difference, though the 1% noise populations are less cooperative in 7 of 8 epochs. The 5% noise populations, however, show a steady and statistically significant increase in their ability to cooperate. It appears that they are learning strategies which allow cooperation in the presence of noise.

The assessment of cooperativeness was performed in a noise-free environment to permit apples-to-apples comparison of the probability of cooperativeness. The bottom panel of Figure 5.3 displays the average fitness of populations, averaged over all 100 replicates. The experiments with 1% noise appear to pay a small fitness penalty for noise events, while the 5% noise experiments pay a larger one. The 5% noise level experiments have a substantial upward trend in fitness over the course of evolution. This is evidence that the strategies evolving under 5% noise are adapting nontrivially to the presence of noise.

Competitiveness and Non-Localized Adaptation   Results for competitiveness are presented in Figure 5.4. Sampling pairs of players as a win/loss Bernoulli variable yields entirely comparable results to placing whole populations into competition as in [31]. The main result is that evolution in the presence of noise yields a competitive advantage. This is in spite of the fact that the arena for evaluating competitiveness is the noise-free environment. The use of two noise level in the experimentation also demonstrates that 5% noise yields a more commanding advantage.

Figure 5.4 displays 64 comparisons representing each possible pair of saved epochs for each of three noise level. Each comparison comprises 10,000 pairwise tests of agents and is made in the form of a confidence interval on the probability that the agent type indexing the row will beat the agent type indexing the column. Agents evolved with a lower noise level index the columns. There are three categories of outcome: row victory, column victory, or a failure to find a significant competitive advantage. These have been gray-scale coded with light gray representing column victory, dark gray representing row victory, and medium gray representing a result in which the confidence interval for probability of row victory includes 0.5.

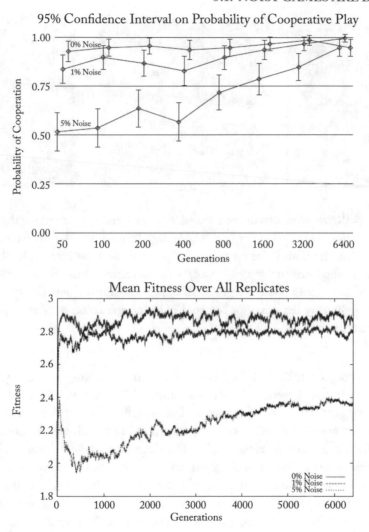

Figure 5.3: 95% confidence intervals on the probability that a population in a given epoch and noise level will play cooperatively in the absence of noise (top) and the mean value over all 100 replicates of the mean population fitness (computed with noise) over the course of evolution.

The gray-scale coding in Figure 5.4 permits the viewer to immediately see that all epochs of agents evolved at 5% noise beat all epochs of agents evolved at 0% noise, in spite of the fact that additional evolution grants competitive advantage when the noise level is not varied. The time factor becomes important in the current experiments in the 0:1 and 1:5 comparisons where agents from a substantially later epoch achieve victory over agents evolved at a higher noise level.

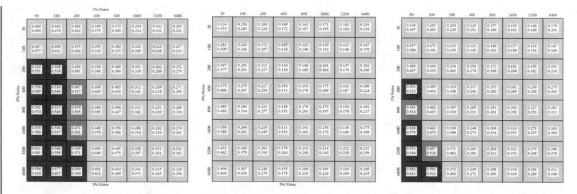

Figure 5.4: Competitiveness results in the form of 95% confidence intervals for the probability that the agents indexing the row will beat the agents indexing the column. The top panel gives results for no noise vs. 1% noise, the middle for no noise vs. 5% noise, and the bottom for 1% noise vs. 5%. Rows are indexed by the lower-noise agent pools with numerical labels for both rows and columns giving the epoch of the agent pool. Dark gray cells indicate significant superiority of the agent-pool indexing the row, light gray the same for the pool indexing the column. Middle gray indicates no significant difference.

One is tempted to conjecture that all three populations are becoming more competitive, and that the agents evolving in the presence of noise are simply doing so faster. The evidence presented in Section 5.2.2 suggests that this is, at best, an oversimplification of the true state of affairs.

The results for non-localized adaptation, shown in Figure 5.6 use the same gray-scale encoding scheme as the competitiveness results. These matrices have a form of skew-symmetry, with entries opposite across the main diagonal being mirror images of one another. This yields 28 distinct comparisons, in the form of pairs of symmetric entries off of the diagonal. However, they are less easy to interpret. The original observation of non-localized adaptation [25] compared two epochs, 100 and 10,000 generations, of a noise free Prisoner's Dilemma structured as a spatial evolutionary algorithm. The current experiments have a much simpler design but report and compare eight epochs rather than two. The hypothesis tested in the initial non-localized adaptation study was that additional evolution grants a competitive advantage even against agent types not encountered before. The initial study supported this hypothesis; the results in [31] and here are consistent, but the new results reveal a more complex situation.

A comparison of two epochs is *consistent* with the non-localized adaptation hypothesis (NLA hypothesis) if either the more-evolved population has a significant advantage or if neither population has a significant advantage. In the zero-noise experiments, summarized in the top panel of Figure 5.6, 26 of the 28 comparisons are consistent with the NLA hypothesis. Both the inconsistent comparisons are on the sub-diagonal representing adjacent epochs. In the experiments with 1% noise, 25 of 28 comparisons are consistent with the NLA hypothesis. The three comparisons that are inconsistent are the comparisons of the three most evolved populations. These three populations also have the greatest temporal separation. The situation grows substantially worse for the NLA hypothesis in the experiments with 5% noise, where only 18 of the 28 comparisons are consistent with the hypothesis. The 5% noise agents sampled at epoch 6,400 are beaten by epochs 200–3,200. This is clear evidence of a substantial decline in competitive ability for the agents evolved with 5% noise at deep time against younger agents evolved at 5% noise.

An explanation for this retrograde non-localized adaptation lies in Figure 5.3. The strategies evolved at 5% noise begin a sharp increase in cooperativeness in the 5th-8th epoch. This corresponds exactly with the sudden decline in competitive ability in the third column, 5th-8th rows of the bottom panel of Figure 5.6. There is an apparent tradeoff between competitive ability and the ability to cooperate in the presence of noise. As they gain the ability to cooperate in a noise environment, the agents evolved in a high-noise environment become more tolerant of (possibly-noise induced) defections. The agents from earlier epochs, presumably less tolerant of defection, exploit this tolerance. This highlights the dilemma of cooperation in a noisy environment: how do we tell a true defection from the noise-induced appearance of defection?

It is worth noting that in the epoch when we hypothesize that creatures with 5% noise are growing more forgiving that Figure 5.5 shows an increase in the fraction of agents with fingerprints close to AllD and (TF2T), both of which are highly non-cooperative strategies. This apparent inconsistency arises from the fact that there are billions of strategies in each tile of the Voronoi tessellation. In order for the populations from later epochs to be more cooperative the agents need only be more forgiving of *their own close relatives*. The competitive advantage enjoyed by agents evolved at 5% noise is consistent with their playing in an AllD or (TF2T) fashion (aggressively) against most other strategies. Within these tiles that are centered on aggressive strategies there are strategies that play very aggressively against most other agents but cooperate with copies or near-copies of themselves. One mechanism for this is handshaking, discussed in Section 5.2.2.

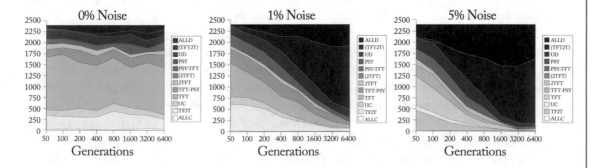

Figure 5.5: The relative size of the contents of the different Voronoi tiles in fingerprint space across the sampling epochs.

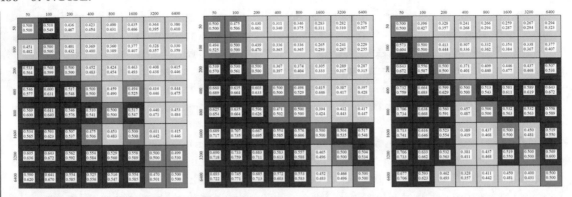

Figure 5.6: Non-localized adaptation results. The top panel gives results for no noise, the middle for 1% noise, and the bottom for 5% noise. Given is a 95% confidence interval on the probability the the agent-pool for the epoch indexing the row will beat the the agent pool for the epoch indexing the column. Dark gray cells indicate significant superiority of the agent-pool indexing the row, light gray the same for the pool indexing the column. Middle gray indicates no significant difference.

The additional tolerance of noise events in the more highly evolved agents from the 5% noise environment does not mean they are becoming pushovers. The between-noise-level comparisons show that the agents are still very strong competitors; the loss of competitive ability in later epochs acts in a purely intramural manner.

Contest Winners and Evolution Dominators The results presented in Sections 5.1.3 and 5.2.2 suggest that it might be worth testing agents sampled from the different noise levels and different populations in a contest setting. The confidence intervals for the mean contest rank of each agent type are displayed, for all epochs, in Figure 5.7. In these tournaments rank 1 is the best rank, and so the better an agent population is the lower it appears in the figure. The tournament ranks are in the range 1–36 while the confidence intervals have a radius of 0.05–0.065 and so are difficult to see.

In the first three epochs, the agents evolved without noise have a significantly better tournament ranking than those evolved with 1% noise which in turn are superior to those with 5% noise. Between the third and fourth epoch agents evolved with 1% noise obtain significantly superior mean contest rank to those evolved with no noise. Between the fourth and fifth epochs the agents evolved with 5% noise also pass the agents evolved without noise. The 1% noise agents retain their superiority to the 5% noise agents throughout all epochs. In [37], it was found that strategies not observed early in evolution became common at deep time (6,400 generations). The results on mean contest ranking echo these results—there is a significant difference in the character of Prisoner's Dilemma agents between early and later evolution.

The ability to win a contest is different from the ability to survive the process of evolution in *a typical* evolutionary algorithm. Winning a tournament involves getting high scores against

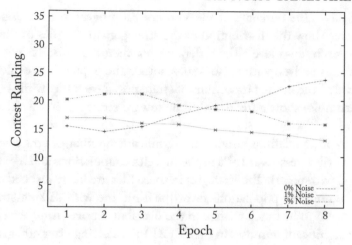

Figure 5.7: 95% confidence intervals on the mean contest rank for agents evolved with each of the three noise levels. These confidence intervals are given separately for each epoch. The confidence intervals are narrow enough to be masked by the impulses marking the data points for each epoch. The horizontal scale is logarithmic: the epochs are exponentially spaced.

a diversity of opponents. Tournament winners do not consistently beat their opponents; they just get relatively high scores while playing them. Evolutionary algorithms select for strategies which get high scores when playing copies of themselves and which beat mutants. We call agents with these two different sorts of competitive ability *contest winners* and *evolution dominators*. A contest winner will receive a high tournament ranking when playing round-robin against a slate of diverse opponents. An evolution dominator will tend to cooperate with agents very similar to itself (its relatives) and beat agents different from itself. These definitions reflect the different objective functions used for scoring contests and for driving the evolution of a small well-mixed population.

It is intuitive that an evolution dominator evolved by a standard (single pool-genetically convergent) evolutionary algorithm is likely to do poorly in a contest unless the contest is packed with strategies similar to itself with which it will cooperate. The evolutionary algorithm used here selects strongly for evolution dominators. It does so less efficiently when noise is present, as the retrograde non-localized adaptation shown in Figure 5.6 demonstrates. The populations evolved with noise are, however, better contest winners. The 1% noise populations are superior to the 5% noise populations. This, in turn, suggests that there is a difference in the contest-winner ability of agents evolved at different noise levels and hence potentially an optimum noise level for such evolution. This level is probably below 5%. These results also suggests that a well-mixed, noiseless, small-population evolutionary algorithm is not an optimal tool for designing entries for a Prisoner's Dilemma contest.

Fingerprint Analysis   The outcome of the Voronoi tile fingerprint analysis is given in Figure 5.5. These figures show the distribution of the strategies in the tiles of the combined populations evolved at each noise level. The difference between the noiseless populations and the two sets of populations evolved with noise is substantial. The populations evolved with no noise change comparatively little in the proportions of strategies in each tile, while both sets of populations evolved with noise show a trend with time toward strategies in the ALLD and (TF2T) tiles.

All three sets of populations contain no significant number of strategies in the ALLC, (2TFT), and PSYC tiles (less than 1.0% in the noiseless populations and less than 0.2% in the populations evolved with noise). The strategies in these tiles are likely just bad strategies. Of the other nine tiles, the noiseless populations favor the TFT tile with 45% of strategies falling in it. There does not seem to be much change in the distribution over time, except that strategies begin to appear in significant numbers in the (TF2T) tile starting after 400 generations (epoch 4). In epoch 1 they account for less than 1% of the strategies; in epoch 4 they account for 2%, and by epoch 8 5% of the strategies are in that tile.

In the populations evolved with 1% noise, there is a significant trend into the ALLD and (TF2T) tiles. The ALLD tile contains 1% of the strategies in epoch 1 and 20% in epoch 8. Likewise, the (TF2T) tile contains 1% of the strategies in epoch 1 and 60% in epoch 8. The trend is even more pronounced in the populations evolved with 5% noise. 1% of the strategies in the first epoch are in the (TF2T) tile and 60% in the last epoch. The move to the ALLD tile happens sooner in the 5% noise populations with 12% of the strategies there in the first epoch and 32% in the last epoch.

The Voronoi tiles do not give information about specific strategies found; they only give information about what other strategies have nearby fingerprints. We also analyzed the occurrences of two specific strategies: ALLD and TFT. In the noiseless populations ALLD never amounts to more than 2% of the total strategies; TFT ranges from 10–22% with a peak in epoch 2. In the 1% noise populations neither strategy is common, they amount to less than 1% of the total, except in epoch 4 in which 3% of the strategies are ALLD. In the 5% noise populations, TFT is also rare (less than 1%), but ALLD exists in significant numbers, ranging from 8–32% of the strategies, peaking in epoch 4.

Another strategy worth analyzing in particular is the Fortress strategy. It was first discovered and analyzed in [37] when it appeared in populations that had evolved for more than 4,096 generations. The three state version of it is pictured in Figure 5.8. Fortress n has n states in its minimal form. It will cooperate indefinitely with another strategy which cooperates with it, but only after a series of $n-1$ defections, and, if the opponent ever defects, it must play the series of defections again before Fortress will cooperate with it. Table 5.1 shows the fraction of strategies with fingerprints close to Fortress 3 for all the populations saved. "Close to" is defined by three different distances: $R = 0.1$ means the strategies are virtually identical to Fortress 3; $R = 1.0$ means they are as similar as strategies in the Voronoi tiles are to their reference strategies, and

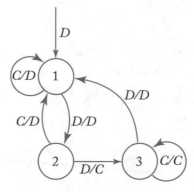

Figure 5.8: The Fortress 3 strategy.

$R = 0.32$ is in between. In the noiseless populations, these strategies are rare, but become more common in the later epochs, peaking at 10% in epoch 8 with $R = 1.0$. The story is radically different in the populations evolved with noise. They also become more common with time, but, instead of being rare, they come to dominate the population with 83% of the strategies within $R = 1.0$ of Fortress 3 in epoch 8 of the 1% noise populations and 86% in the 5% noise populations.

5.1.4   CONCLUSIONS AND NEXT STEPS

This chapter introduces results on competitiveness, non-local adaptation, and probability of co-operative behavior in the presence of noise to reach the following conclusions. First, employing noise in a finite state representation increases the fraction of states that is under selection. Second, having a higher fraction of the genome under selection enhances the efficiency of evolutionary training.

The work presented here also highlights two distinct forms of competitive ability, the ability to dominate an evolving population and the ability to win a contest against a diverse selection of opponents. These two abilities, while not opposites, are demonstrated in this study to have some degree of trade-off. In particular, we conclude that a well-mixed, small-population evolutionary algorithm will favor the production of evolution dominators over contest winners. We also note that evolving with noise enhances the algorithm's ability to produce contest winners and that adding 1% noise does this to a greater degree than adding 5% noise. This suggests that the noise level can be optimized for the production of contest winners, a possible next step for this research.

Results on the number of transient and unused states in the three different agent types suggests that the more efficient evolution enjoyed by agents evolved with noise may have permitted evolution to exert a greater parsimony pressure on the size of the active genome in those agents. The genome size, 8 states or 68 bits, used in this study leaves only modest room for

Table 5.1: Fraction of saved population within various radii of the fingerprint of the strategy Fortress-3 for different epochs and noise levels.

Epoch	R = 0.1	R = 0.32	R = 1.0
0% Noise			
50	0.0000	0.0004	0.0471
100	0.0000	0.0000	0.0271
200	0.0000	0.0025	0.0354
400	0.0000	0.0108	0.0387
800	0.0000	0.0000	0.0046
1600	0.0308	0.0308	0.0446
3200	0.0196	0.0283	0.0867
6400	0.0413	0.0337	0.1008
1% Noise			
50	0.0000	0.0004	0.0171
100	0.0175	0.0238	0.0425
200	0.0646	0.0671	0.1083
400	0.1179	0.1471	0.2525
800	0.1988	0.2046	0.3588
1600	0.3125	0.3350	0.5483
3200	0.4646	0.4846	0.7396
6400	0.5729	0.5742	0.8329
5% Noise			
50	0.0000	0.0000	0.0342
100	0.0088	0.0092	0.0721
200	0.0683	0.0742	0.1371
400	0.1988	0.2112	0.3162
800	0.2658	0.2804	0.3646
1600	0.3917	0.4238	05529
3200	0.5575	0.5642	0.7183
6400	0.6775	0.6775	0.8625

parsimony pressure to act. A follow-up study with more states could be used better understand the interaction of noise and evolution to affect state usage.

It is demonstrated that the non-localized adaptation hypothesis from [25] is too simple. While additional evolution often conveys a competitive advantage, it doesn't do so invariably. This study opens a number of doors for understanding why this is so. The delineation of distinct types of competitiveness (evolution dominator and contest winner) also permits a more nuanced phrasing of the non-localized adaptation hypothesis.

The hypothesis that the fraction of the genome under selection is an important variable deserves additional study. The work presented here uses eight-state Mealy automata as its representation. This representation encodes a space of $2^{68} = 2.95E10^{20}$ distinct automata (many of which encode the same strategy), but each of these represents only 68 bits of information. It seems likely that the character of evolution in the presence of noise may change significantly if the genome size of agents is increased or permitted to evolve. In addition, the finite state representation is adept at failing to use portions of its genome. The representation used in this study, directly encoded finite state machines, was one of twelve used in [14, 21]. The genome-fraction question could be addressed in the context of any of these representations and many others.

The questions about genome size can be asked in many contexts well beyond the Prisoner's Dilemma. Fingerprinting may be employed for any simultaneous two-player game with a finite number of moves [76]. Many of the results supporting the results in this study are restricted to games with two moves (the mathematics becomes more complex when three or more moves are available).

5.2   EVOLUTIONARY VELOCITY

One advantage of fingerprints is to place a metric-space structure on the space of Prisoner's Dilemma strategies. Fingerprinting was used in [37], with a finite state representation, to demonstrate that the strategies that arise have different distributions for different population sizes and in different epochs. The latter result, that strategies rare or absent at the beginning of evolution become common after thousands of generations of evolution, was surprising. In Section 2.3, fingerprints were used to demonstrate that the rate of appearance of several well-known strategies varied between a direct finite state representation for Prisoner's Dilemma playing agents, a cellular representation for finite state agents, and a new type of representation called a function stack, a modified form of Cartesian Genetic Programming. In this section, the metric space structure induced by fingerprints is used to compute a new quantity: evolutionary velocity. The evolutionary velocity of a population is the distance that its mean fingerprint moves in one generation. The experiment explored in Section 5.1 is analyzed with the evolutionary velocity in this section.

5.2.1   DEFINITION OF EVOLUTIONARY VELOCITY

A *velocity* is the change of a quantity over time. The natural measure of time in the evolutionary algorithms used in this section is generations. The fingerprints of Prisoner's Dilemma playing agents used in this section map the strategies to points in 25-dimensional Euclidean space. A population may then be said to have a mean position in fingerprint space in each generation.

Definition 5.1   The *evolutionary velocity* of an evolving population in a given generation is defined to be the change in the position of the mean fingerprint of the population between the last generation and the current one using the standard Euclidean metric.

The small, well-mixed populations used in this chapter ensure that a population will move to and stay near a state of relatively low diversity. When a population is relatively non-diverse, the fingerprints will be close together. A common event in an evolving population of game-playing agents is a succession event. This happens when a new agent type appears that can exploit a flaw in the current dominant agent type. The population diversity briefly increases as the new agent type takes over. There should be some succession events visible as simultaneous changes in the population average score and spikes in the evolutionary velocity.

Based on earlier work [31] we know that being exposed to noise yields a competitive advantage to Prisoner's Dilemma playing agents, an observation confirmed via a different evaluation of competitive ability in Section 5.1.3 of this study. Because of this we conjectured that evolutionary velocity would be lowest in the noise-free environment and highest in the environment where agents were evolved with the most noise. In Section 5.2.2 we test and discuss this incorrect conjecture.

5.2.2   ANALYSIS OF EVOLUTIONARY VELOCITY

Fingerprint-based evolutionary velocity is introduced here. When small, well-mixed populations of Prisoner's Dilemma playing agents are trained with an evolutionary algorithm the populations rapidly become non-diverse. This means that, most of the time, a single agent type with minor variation dominates the population. An important event is the *succession* in which a new agent type, able to out compete the current dominant type, appears and takes over the population. Succession events are sometimes accompanied by sudden changes in the population mean fitness. In other cases the fitness difference is relatively small and the succession is difficult to detect. One potential application of evolutionary velocity is to detect such events.

This type of detection is prototyped in this study by displaying fitness and evolutionary velocity on the same set of axes. One representative population from each noise level are displayed in this fashion in Figure 5.9. For all three noise levels we note that abrupt changes in average score are often accompanied by pronounced spikes in evolutionary velocity. The size of the changes in fitness is not well correlated with the size of the evolutionary velocity spikes. It is intuitive that these sizes need not be correlated in magnitude. A relatively modest shift in strategy, and hence fingerprint, can result in a large change in score and vice versa.

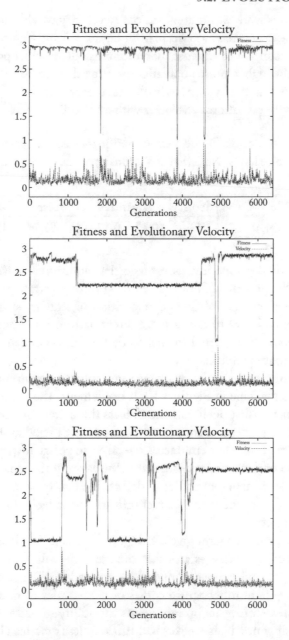

Figure 5.9: This figure displays overlaid plots of fitness and evolutionary velocity on the same axes. Each of the three panels are results for one replicate. The top panel is from a population evolved with 0% noise, the middle a 1% noise population, and the bottom 5% noise. These plots are illustrative rather than representative.

A hypothesis of this work was that evolutionary velocity would increase with noise level. That simple hypothesis explains the patterns in both Figure 5.5 and Figure 5.6. Using both these indicators it is not unreasonable to suppose that all three collections of populations are heading to the same place with the higher-noise populations getting there faster. The confidence intervals in Table 5.2 demonstrate that the hypothesis is the exact reverse of the true state of affairs. Noise is inversely correlated with evolutionary velocity with all the differences statistically significant.

Table 5.2: 95% confidence intervals for the mean evolutionary velocity over all generations and replicates for three different noise levels. Note that evolutionary velocity is *inversely* correlated with noise level.

Noise	Mean EV	Noise	Mean EV	Noise	Mean EV
0%	(0.164, 0.168)	1%	(0.134, 0.138)	5%	(0.125, 0.129)

Given the substantial competitive advantage that additional noise grants, this suggests that agents evolved in the presence of noise are moving through strategy space more efficiently. Evolutionary velocity is a measure of change of position in strategy space. If there is a reason the higher noise agents will tend to take a more direct path to good positions in the adaptive landscape, then all the results reported in this study remain consistent. Such an explanation follows with some necessary preliminaries.

One feature of fingerprints is that they measure only the asymptotic behavior of agents. Examine the finite state machine in Figure 5.10. It implements the strategy *vengeful* which cooperates until its opponent's first defection and defects thereafter. In a noise-free environment this is a fairly competitive strategy. it cooperates perfectly with itself and defends itself aggressively against anything that does not embrace a no-first-use policy toward defection. The noise used in the computation of fingerprints ensures that the asymptotic behavior and hence the fingerprint of vengeful is identical to that of always defect. The ability of vengeful to cooperate with copies of itself are strongly dependent on perfect information: a lack of noise. Its cooperation is embedded in transient states.

The implementation of vengeful given in Figure 5.10 is a simple example of the dependence of success on transient states—states that the machine will leave and never return to if the opponent makes a particular move. Populations evolved in the absence of noise are likely to exploit such transient states to recognize copies or near-copies of themselves. This recognition strategy is called *handshaking* and grants agents the ability to preferentially cooperate with strategies that share their genes. Noise makes handshaking less practical, because the handshake must be able to tolerate noise. The practicality of handshaking also drops as the level of noise increases both because the imperative to correct for noise grows and because the potential for more complex patterns of noise grows. A strategy that can deal with one noise event and still handshake cannot deal as easily with two. Given that the agents used in this study have eight states, there may not be room in the genome to handle multiple-noise errors in a handshake.

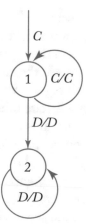

Figure 5.10: The finite state diagram for a two-state implementation of the strategy *vengeful*.

Consider now a population of agents evolved without noise. There is evolutionary pressure to make the portion of the state-diagram used to compete with copies of yourself small. This reduces the probability it will be modified by mutation and so makes it more heritable. This secondary evolutionary pressure to keep the active genome small is a form of *parsimony*. It is also advantageous to have this portion of the state diagram immediately accessible from the starting state of the transition diagram. This evolutionary pressure thus favors the use of "early" states in the transition to manage the interaction of an agent with its close relatives. Such states are often transient and thus not measured by fingerprinting. "Later" states might be rarely used in fitness evaluation. Mutation, thus, would make them more or less random. We deduce that agents evolved without noise will have evolutionary velocity considerably closer to the fundamental random-walk velocity that would result from applying mutations to a population with random selection than would a population evolved with noise. We offer this as a tentative explanation for the higher evolutionary velocity of the zero-noise populations. We turn now to the question of more efficient travel though strategy space by agents evolved at higher noise levels.

We have already discussed why an agent evolved in the presence of noise will find it harder to depend on transient states. In fact, any state accessible from the starting state has a positive probability of being reached when noise is present. This means that the number of states used during a fitness evaluation will be higher when noise is present and, to a degree, the evenness of access to states will increase with noise. This means that agents evolved in the presence of noise have a higher fraction of their genetic material subjected to selection pressure. It is intuitive that this increase in the generality of selection pressure across the genome will grow with noise, at least for "low" noise levels.

At this point a thought experiment may be valuable. A population of agents evolved in the absence of noise inherit many features in their transition diagrams that are not subjected to selection pressure. A mutation will, with positive probability, place these untested features into the portion of the transition diagram that is relevant to selection. Likewise, mutation can detach parts of the transition diagram from those portions undergoing selection and, once they are detached, re-randomize them. This means the rate at which a population develops, and later expresses, non-adaptive features is much higher when the fraction of the genome undergoing selection is lower. We have already demonstrated that the fraction of the genome undergoing selection is higher in the presence of noise. This means that evolutionary velocity generated by the exploitation of and elimination of non-adaptive features resulting from the long-term action of mutation will be higher as noise levels drop and will be highest at zero noise.

The agents evolved with noise are, in essence, making far better use of their fitness evaluations. Since a greater fraction of their genomes are under test, they are more likely to retain adaptive behaviors once those behaviors are acquired. They are less likely to gain non-adaptive features that, once placed under selection pressure by a mutation, must be eliminated from the gene-pool. We conclude that agents evolved in the presence of noise move toward good locations (those with a competitive advantage over a larger set of opponents) in the adaptive landscape more efficiently than agents evolved without noise. To be clear: no-noise agents have a higher fingerprint-based evolutionary velocity that generates a lower rate of acquisition of adaptive behavior. This results from the lower proportion of their genome that is under selection in any given generation.

To test this reasoning, we analyzed the state use of the agents in the experiments. The results of this analysis are shown in Figure 5.11. This figure shows that the noiseless populations use significantly more transient states. It also shows that, in the early epochs (the first five for 1% noise, the first three for 5% noise) the agents evolved with noise use more of their states than those evolved without noise. In later epochs the number of states used by agents evolved with noise decreases. We conjecture that this is driven by the increase in evolutionary efficiency. This evolutionary efficiency has two sources: a lower rate of discarding parsimony gains and a more frequent testing of states within each finite state machine.

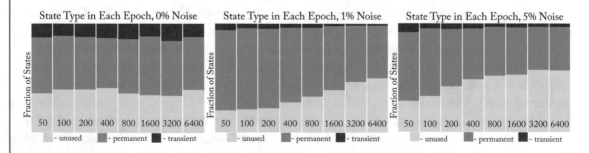

Figure 5.11: Fraction of states, across all members of elite populations within an epoch, that are transient, permanent, or unused. The top panel shows the result for no noise, the middle for 1% noise, and the bottom for 5% noise.

All the evolving agents are subject to the same parsimony pressure to have smaller (i.e., more heritable) genomes. If the agents evolved with noise are evolving more efficiently and wasting less time weeding out non-adaptive features, then the parsimony pressure acts on them more strongly. The adaptive quality of a larger fraction of the states accessible from the starting state of a machine are testing in each fitness evaluation because of noise events. This is the first source of evolutionary efficiency. The second source of efficiency is as follows.

The results of parsimony pressure may be set aside during a succession event. A mutant able to out compete the current dominant strategy is likely to be using states reconnected by the mutation. The zero noise agents, which run around the adaptive landscape at a higher rate, are likely to have a higher rate of succession events and so lose parsimony gains more often. This tentative explanation for the smaller number of states used by agents evolved in the presence of noise suggests that experiments using agents with more states might provide valuable additional data.

CHAPTER 6

Describing and Designing Representations

The central theme of this book is that representation matters. Consider the mixed strategy representations common in classical game theory—probability distributions on the set of moves for a game. This is a simple representation with the cardinal advantage that many aspects of it can be solved mathematically. As a representation, it also runs afoul of George E. P. Box's dictum that all models are wrong, but some models are useful. The obvious way in which mixed models are wrong, as a representation for agents that simulate human beings, is that they simplify most of the qualities of humans that a researcher might want to model right out the modeling process. There is a problem in that models are considered acceptable in classical game theory only if they can be solved. With powerful computers available at nugatory prices, agent-based simulation is a viable and potentially valuable alternative to traditional theorem-and-proof methods.

On the other hand, the material presented thus far in this book demonstrates that agent based modeling is a technology fraught with peril. The traditionalists are right to be cautious about moving beyond models that can be addressed via pure mathematics. While such models are harshly limited in the scope of what they can model and predict, they enjoy the absolute certainty available only from formal, mathematical proof. Agent-based simulations immediately invoke the potential for unintended consequences and chaos, in both the colloquial and formal mathematical senses.

The motivation for this book is to make a start at reducing the problems with incomplete control and specification of agent based models by identifying them. This chapter forms a list of the qualities that the author and her collaborators have identified as requiring control in the experimental design of agent-based studies. These qualities also form a checklist or partial design space for agents. These qualities are internal and external state information, the ability to distinguish other individuals, the ability to learn, and access to random numbers. In the last section of the chapter the issue of complex representations is raised. These representations have greater potential for modeling behavior but with correspondingly greater design and control problems.

6.1   DOES THE REPRESENTATION HAVE INTERNAL STATES?

Internal state can be formally defined as the ability to react differently to the same external situation when encountering it again. The various finite state representations presented thus far clearly have state information. They are possessed of explicit internal states and have a register that remembers which state they are in. Internal state, however, can be implicit. Let's look at a representation too simple to be included in Chapter 2.

Suppose we are playing 150 rounds of IPD and so need to specify 150 moves. The representation consists of a string of thirty characters, **C** or **D**. To generate the 150 moves, cycle through the thirty characters five times. This is an example of a *string representation*. This representation does yield 150 moves, but the agent specified does not react in any way to the opponent's actions. When a representation does not base its moves on what it's opponent does it is called a *non-reactive* representation. Otherwise the representation is *reactive* . In general, non-reactive representations do poorly. This representation is put forward because it *does* have state information. The state is stored as the current position in the string as it is cycled through.

Checking the formal definition of having state information, imagine that the agent whose thirty moves are *CDDCD CDCCD DDDCC CDCCC DDDDD CDCCD* is playing an agent whose strategy is "always cooperate." Then on some moves it defects and on others it cooperates, even though the situation is always the same: my opponent always cooperates. This shows it can react differently to the same situation and so does have state information. When a representation has state information we call it *state conditioned*. A strategy that is not state conditioned but which is reactive is called *purely reactive*.

The following list classifies the representations from Chapter 2 as to their type, relative to the features of being reactive and state conditioned. It also specifies, for the state conditioned strategies, how the internal state information is stored.

Finite State Machines. These are reactive, state conditioned strategies that store their state information in an explicit register that records their current internal state.

Boolean Formulas (no delay line). Boolean formulas are purely reactive strategies.

Boolean Formulas (delay line). Boolean formulas with a delay line are reactive, state conditioned strategies that store their state information in the Boolean variables used by the delay lines.

Function Stacks. These are reactive, state conditioned strategies that store their state information in the registers in each node that retain the output value from the most recent evaluation of the node.

ISAc Lists. These linear genetic programming strategies are reactive, state conditioned strategies that store their state information in their instruction pointer. This is similar to the position-in-the-string state in the example above.

Look-Up Tables. These are purely reactive strategies—but they use external state information; see Section 6.2.

Markov Chains. These are purely reactive strategies—but they use external state information, see Section 6.2, and they have access to random numbers, see Section 6.4.

Artificial Neural Nets. These are purely reactive strategies; this is not an intrinsic quality of artificial neural nets, the ones used in Chapter 2 are *feed forward* but there is another sort called *recurrent nets* which do retain state information [66].

Only one sort of widely published agent representation is not reactive—these are the mixed strategies of classical game theory. Agents that capture non-trivial aspects of human behavior need to be reactive. Both state conditioned and purely reactive representations appear in the literature with the choice to use state information being a critical design decision.

More Complex Internal States   The representations classified as having or not having internal state information in this section have only one sort of internal state, unless one regards the many bits stored in individual logic trees or in the state-last-time registers of the nodes of a function stack. In published analysis the overall state of these bits is taken to be a single master state. This simplifies analysis without losing any information and these different registers holding state information have identical type. In Section 6.5 we will look at the potential for having multiple, interacting types of states.

6.2   DOES THE REPRESENTATION USE EXTERNAL STATE INFORMATION?

In Chapter 3, one of the factors identified as affecting the agent training process was the *memory depth* of the agents—the number of past actions, their own and their opponents, that they had access to. This represents a type of state information that exists outside the agent. The impact of memory depth—together with a novel variation operator that modifies memory depth—are explored in [80]. An extension of this work that also includes geography appears in [74]. An interesting result in this research is that, when memory depth can grow over the course of evolution, agents are discovered that do not appear when evolution starts with full-memory depth strategies.

Representations that use tags, like the ones in Section 3.4, have the ability to recognize the type of agent they are interacting with. This is another type of external state information. This sort of tag can be specific enough to permit agents to be recognized as individuals or, and this is more common, only designate agent types.

In [101], the standard Prisoner's Dilemma experiment was modified by permitting agents to choose their opponents. This choice was conditioned on a decaying moving average of the scores resulting from past plays. This moving average is a scalar, numerical summary of past play. This number functioned in much the same way that tags do. The ability to choose which other agents to play (and to refuse offers of play from those that do not meet the standard) also cause social networks to form among the agents.

Any features in the simulated environment where the agents interact can serve as external state information. The more details the simulation incorporates, the more external state information is available to the agents. When an agent simulation incorporates some sort of geography, the relative position of agents can be used as another form of external state information.

Ant colony optimization [55] uses external state information in the form of simulated pheromones to permit *stigmergy*. The definition of stigmergy is a consensus social network mechanism of indirect coordination, through the environment, between agents or actions. Natural ants leave scents—which evaporate over time—to lead other ants to food sources, the organizing principle behind ant trails.

While stygmergic self-organization is often thought of as providing organization in a spatial environment it could be used in a variety of ways. Imagine a decaying numerical register, accessible as external state information to all agents, that is incremented whenever any player receives the sucker payoff; this would be a general signal to an entire population that an agent was being ripped off. This sort of signaling represents a largely unexplored and interesting direction for game-playing agents.

Another sort of evolutionary system that employs external state information is particle swarm optimization (PSO)[75]. This system uses particles with momentum that travel through the search space, examining the objective value of a point being optimized. The particles, in addition to following their own momentum, bias the direction of their search toward the best result they have found and the best result any particle has found. Vectors in the directions of these two "best" positions, randomly scaled, are added into the momentum vector.

The use of the two best-result reference points prevents the particles in the swarm from simply moving in a line and reporting a transect of the search space. They also organize the search for good points as a social system. The PSO system uses several sorts of external state information to modify its basic Newtonian dynamic search: position, momentum, direction of personal best result, direction of global best result. This sort of external state information might be transferable to game playing—but there is not an obvious way to do it.

Cultural algorithms [94] are a type of representation build around external state information. While originally applied to function optimization, several forms of cultural algorithm have been created and adapted to tasks as diverse as car driving algorithms and modeling the behavior of Pleistocene caribou. The algorithm is similar to a traditional evolutionary algorithm but the agents are augmented with connections to a belief space of hypothesis about the problem they are solving or how search of the problem space should be conducted.

While much of the research on creating plausible and expressive game playing agents has concentrated on representation with internal state information, agents that model human behavior will require a plausible environment, capturing critical features of the real world. Such an environment will be composed of external state information, suggesting that an urgent area for additional work is the study of game-playing agents that use and exploit environmental features. The only part of this that is well started is the research into the use of tags.

6.3 CAN THE REPRESENTATION LEARN?

The representations presented in this book do not learn; each represents a fixed (if possibly adaptive) strategy for playing a game. Evolutionary algorithms are used to train the agents—so in a sense we are using evolutionary algorithm as a learning algorithm to teach agents to play a game well. This point of view is at odds with the more Darwinian view that we are destroying some agents and generating variations of others, a process which appears to have no learning in it at all, only evolution. These antithetical views are both correct descriptions of the agent training algorithm. The apparent paradox is resolved by noting that learning is taking place, but at the population level rather than the individual level.

The issue of learning is sometimes contentious in evolutionary computation because the definition of learning is not simple. Evolution is thought to be a process where adaptation takes place by the reassortment and modification of genes. When recent biological discoveries like gene regulatory systems are included in the description, evolution becomes a system for creating complex adaptive agents that can pull many, many phenotypes out of one genotype. This makes *developmental biology* a type of learning, in which interactions between the organism and its environment determine which genes lie dormant, which genes are expressed, and how those genes are expressed. Simulated developmental biology has been implemented to extend the reach of evolutionary computation [68].

There is an entire field of learning algorithms, beyond the scope of this text, but it is worth noting that *mimetic algorithms* [45] are evolutionary algorithms in which the starting point for a learning algorithm is evolved, but the fitness of a member of the evolving population is the quality of the outcome of the learning algorithm. Mimetic algorithms, also called Baldwinian evolutionary algorithms, are hybrids of evolutionary algorithms with learning algorithms.

A simple reason why such algorithms are valuable lies in the fact that evolutionary algorithms are better at exploring for basins of attraction in the search space containing new optima than at finding the peak of an optima, once located. Any of a number of hill climbing algorithms have the opposite property—they excel at finding peaks. A mimetic algorithm that evolves starting points for hill climbers is a powerful type of optimizer.

This simple example does not transfer directly to the training of game playing agents. The co-evolutionary nature of the training of game playing agents means that there are no fixed hills to climb, but other sorts of mimetic algorithms could be implemented. Consider the Markov chain representation from Chapter 2. An agent would learn by adjusting its own probabilities of cooperation. This would be done randomly and, if the agent's score did not get worse, the change would be retained, otherwise it would be discarded. This sort of trial-and-error adaptation is a perfectly good learning algorithm and, so far as the author knows, has not been tried yet for game playing agents.

It is indisputable that human beings learn, at least sometimes, and so a system for modeling human behavior should have at least the option of incorporating learning. This is not to say that learning must always be included. Einstein's observation that everything should be made as simple as possible but no simpler applies here. The researcher should consider if learning is needed for the situation being modeled.

6.4   DOES THE REPRESENTATION USE RANDOM NUMBERS?

Figure 6.1 shows the way that the strategy spaces encoded by several of the representations presented in this book contain one another as sets. It is worth noting that the innermost set—consisting of three purely reactive strategies—include the representation that demonstrated the lowest level of cooperativeness (neural nets) and one of the most cooperative (look-up tables). To make matters worse, logic trees exhibited an intermediate level of cooperation. A striking feature of this diagram is that the Markov chains, while they generalize look-up tables, are completely outside of all the other representations. This is because they are the only representation that incorporates *randomness*.

The representation for Markov chains is very similar to that for look-up tables—external state information in the form of the plays of the agent and its opponent are used to index a vector of responses. The difference is that, while the look-up table has a "C" or a "D" as its response, the Markov chain has a probability of cooperation. This means that, when it plays IPD, an agent encoded as a Markov chain must throw random numbers. None of the other representations use random numbers—they all encode deterministic strategies.

In Chapter 5, randomness in the form of noise was introduced into the IPD simulation, but this is different. While there is randomness, it appears in the communication between agents, not in the internal organization of individual agents. Given that the Markov chains are, in effect, probabilistic look-up tables, it is not hard to see ways of making probabilistic modifications of all the other agent styles. This creates a titanic collection of additional representations; it is not clear which modeling problems might benefit from incorporating randomness in this fashion.

In retrospect, the realization that Markov chains are very special and different was too slow. The stepwise path—by changing the number of available probability levels—from look-up tables to Markov chains with unrestricted probability levels, yielded surprising results. The number of available probability levels had a substantial impact on the agent's behavior.

The number of possible experiments that could be performed by granting agents access to random numbers is huge. Even the basic Markov chain experiments have many unexplored avenues. Suppose, for example, that we give the agents a small number of probabilities but permit

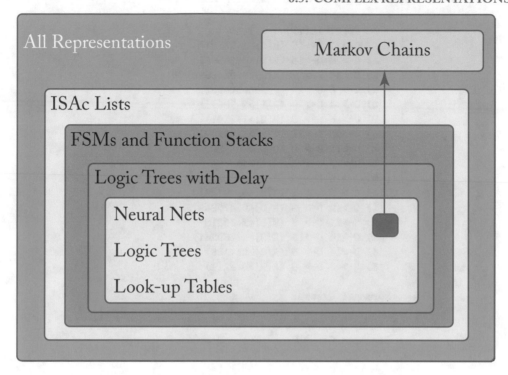

Figure 6.1: The containment diagram for the sets of strategies encoded by a number of different representations.

those agents to evolve those probabilities. This is a novel agent representation and might enable discoveries about what collections of probabilities grant a competitive advantage in various situations.

6.5 COMPLEX REPRESENTATIONS

Figure 6.2 shows an instance of a complex representation that incorporates two finite state machines. Experiments with this agent representation appear in [22]. The upper machine is driven by the outcome of Boolean tests. These tests are one of a comparison of a numerical register with a constant or a statement about the opponent's last action. The numerical registersused to drive this machine are maintained by the second FSM.

The second finite state machine is the agent's evolvable, simulated hormonal system. It has four states and starts in state 0. This machine is driven by the outcome of the simultaneous plays of the two agents, labeled by the PD payoffs S, D, C, or T, that the agent receives. If the agent is in state 0, for example, and receives a **D** payoff it transitions to state 2. Each time the agent enters a state, it increments the numerical register associated with that state by 1. After

```
Start: D->10
        If(F) If(T)      Test
   0) D-> 0 C->15 (R[0]>1.23131)
   1) D-> 4 D->14 (R[2]<36.4598)
   2) D->14 D-> 9 (R[2]<0.0418594)
   3) C-> 8 C->14 (Opponent Def.)
   4) D-> 4 D-> 9 (R[1]>9.28591)
   5) D-> 5 D-> 3 (R[2]>11.141)
   6) C->12 C->15 (R[1]>1.11704)
   7) D->13 D-> 0 (Opponent Def.)
   8) D-> 7 C-> 6 (R[1]>12.7847)
   9) D-> 5 C-> 5 (R[3]>2.60719)
  10) C-> 8 D-> 1 (R[0]<43.9427)
  11) C->12 D-> 9 (R[0]<0.472249)
  12) C->11 C-> 6 (R[3]<1.78318)
  13) C->14 D->15 (R[2]>0.530241)
  14) D->14 C-> 5 (R[0]>21.139)
  15) D-> 3 D-> 8 (R[2]>8.2214)


Hormonal System:


               (Score Recieved)
   Register C    S    T    D
      R[0]  0    1    3    2
      R[1]  0    0    3    1
      R[2]  0    2    1    3
      R[3]  3    0    2    0
```

Figure 6.2: An example of an evolved hormonal side effect machine. The finite state play controller initially defects and begins in state 10. The hormonal system's side effect machine always starts in state 0 and its transitions are driven by the score the agent receives.

each play of Prisoner's Dilemma, all the numerical registers are multiplied by a decay factor of $\omega = 0.95$. Examples of the numerical levels of hormones over the course of a 150 round session of IPD are shown in Figure 6.3.

The hormonal traces of four different agents are shown in Figure 6.3, all playing against the same opponent. This demonstrates that these agents all developed very different hormonal systems. The practical impact is that the hormonal information gives the agent memory that spans the entire history of play thus far, favoring more recent events because of the decay term. The presence of this lingering memory of past events gives the agent the potential for greater adaptability.

The ability of the agents to condition the transitions of their primary finite state machines on their opponent's last action leaves open the option for evolution to discard the hormonal information and regress to the FSM representation used in Chapter 2. Instead of doing this,

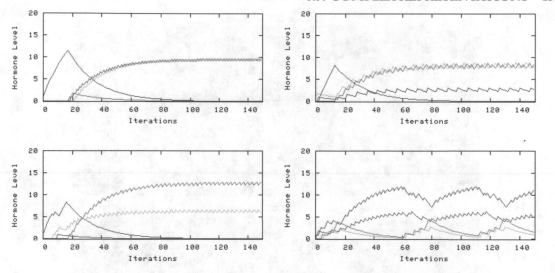

Figure 6.3: Examples of the level of hormonal registers of agents evolved with a round-robin fitness function over 150 Iterations of Prisoner's Dilemma. The values of the hormone levels were sampled while the agents were playing against a agent with the Pavlov strategy.

the agents used the register comparisons (as opposed to information about their opponent's last actions) at a rate significantly higher that would be explained by random chance.

The *meaning* of the hormones evolved by these agents is obscure. The price of permitting the agents to select their own hormonal systems via evolution is that evolution does not have any respect for the human desire to have a comprehensible solution to problems. This is, in a way, a flaw in this representation. Its complexity defies simple analysis.

Once we have specified a complex representation, the next natural question is, did this representation behave differently from the more basic representations. A natural comparison to make for the hormonal agents is with finite state agents that condition on their opponent's last action. Such a comparison is shown in Figure 6.4. The baseline agents segregate neatly into mostly cooperative populations and a few that fall into the always defect Nash equilibria; a few populations change behavior part way through. The hormonal agents, in contrast, are almost 60% in the always defect Nash equilibria, but the remaining agents exhibit a far more diverse variety of behaviors.

The gray middle region of the lower panel of Figure 6.4 is almost absent in the baseline agents. This score region is characteristic of complex play of the IPD. The hormonal agents are in the middle of the pack, compared to the various representations in Chapter 2 for how cooperative they are. Placed in competition with non-hormonal agents, hormonal agents exhibit a marked competitive advantage. The surface of this perhaps overly complex representation has just been scratched.

Non-Hormonal Agents

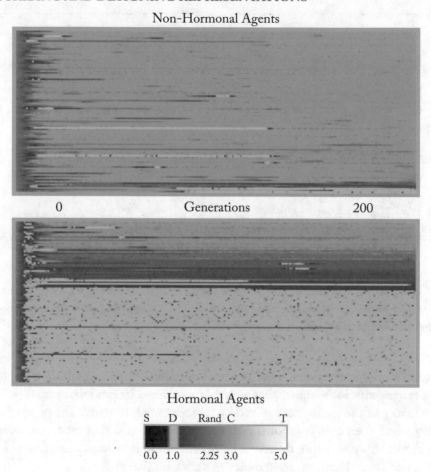

Figure 6.4: The mean population scores for 100 populations of agents evolved for 200 generations using a round-robin fitness evaluation. The upper panel shows non-hormonal agents while the lower panel shows agents with active hormonal systems. Color coding for scores is given by the bar at the bottom of the figure; Rand (shown in red) is the score a random player gets against itself. The populations are sorted by their mean score in the final generation with the highest score at the top.

The paper [22] on the hormonal agents was the third and last in the series on this sort of complex agent system. Earlier papers looked at augmenting neural nets with hormonal inputs and at single, fixed hormonal variables, such as the Boolean value "has my opponent received more **T** payoffs than I have." This very simple piece of long term information detects predatory opponents and so was a natural choice.

While this work on complex representations case in progress by the author's collaborators, the case that even simpler representations were not subject to proper controls kept becoming clearer. Continuation of the complex agent project waits on a better understanding of the issue raised in this book. Certainly these complex representations show promise as versatile agent models, but without a better understanding of the control structure of agent-based game-playing agents, it will be impossible to tell if the agents exhibit desired behaviors because the simulation went well or because of poorly understood factors in the simulation design.

Bibliography

[1] N. Anbarci. Divide the dollar game revisited. *Theory and Decision*, 50:295–304, 2001. DOI: 10.1023/A:1010363409312. 102

[2] D. Ashlock. GP-automata for dividing the dollar. In *Proc. of the Genetic Programming Conference*, pages 18–26, Cambridge, MA, MIT Press, 1997. 102, 107, 112

[3] D. Ashlock. *Evolutionary Computation for Optimization and Modeling*. Springer, New York, 2006. DOI: 10.1007/0-387-31909-3. 4

[4] D. Ashlock. Training function stacks to play iterated prisoner's dilemma. In *Proc. of the IEEE Symposium on Computational Intelligence in Games*, pages 111–118, 2006. DOI: 10.1109/cig.2006.311689. 47, 62, 63

[5] D. Ashlock. Cooperation in prisoner's dilemma on graphs. In *Proc. of the 2005 IEEE Symposium on Computational Intelligence in Games*, pages 48–55, 2007. DOI: 10.1109/cig.2007.368078. 24

[6] D. Ashlock and A. Aherk. Non-local adaptation of artificial predators and prey. In *Proc. of the Congress on Evolutionary Computation*, volume 1, pages 41–48, Piscataway, NJ, IEEE Press, 2005. DOI: 10.1109/cec.2005.1554665. 93, 127

[7] D. Ashlock, W. Ashlock, S. Samothrakis, and S. Lucas. From competition to cooperation: Co-evolution in a rewards continuum. In *Proc. of the IEEE Conference on Computational Intelligence in Games*, pages 33–40, 2012. DOI: 10.1109/cig.2012.6374135. 44, 48, 65, 72

[8] D. Ashlock, W. Ashlock, and G. Umphry. An exploration of differential utility in iterated prisoner's dilemma. In *Proc. of the 2005 IEEE Symposium on Computational Intelligence in Bioinformatics and Computational Biology*, pages 271–278, 2006. DOI: 10.1109/cibcb.2006.330946. 24, 63

[9] D. Ashlock, E. Blankenship, and J. D. Gandrud. A note on general adaptation in populations of painting robots. In *Proc. of the Congress on Evolutionary Computation*, pages 46–53, Piscataway, NJ, IEEE Press, 2003. DOI: 10.1109/cec.2003.1299555. 75, 93, 127

[10] D. Ashlock and K. M. Bryden. Non-local adaptation in bidding agents. In *Smart Engineering System Design: Neural Networks, Evolutionary Programming, and Artificial Life*, volume 15, pages 193–199, ASME Press, 2005. 127

[11] D. Ashlock, K. M. Bryden, S. Corns, and J. Schonfeld. An updated taxonomy of evolutionary computation problems using graph-based evolutionary algorithms. In *Proc. of the Congress on Evolutionary Computation*, pages 403–410, Piscataway, NJ, IEEE Press, 2006. DOI: 10.1109/cec.2006.1688295. 35

[12] D. Ashlock, S. Gillis, and G. Fogel. Ring optimization with extinction. In *Proc. of the Congress on Evolutionary Computation*, pages 1311–1318, 2015. DOI: 10.1109/cec.2015.7257040. 50

[13] D. Ashlock, S. Gillis, J. Garner, and G. Fogel. Evolving dna classifiers with extinction based ring optimization. In *Proc. of the IEEE Conference on Computational Intelligence in Bioioinformatics and Comutational Biology*, pages 1–8, 2015. DOI: 10.1109/cibcb.2015.7300312. 50

[14] D. Ashlock and E. Y. Kim. The impact of cellular representation on finite state agents for prisoner's dilemma. In *Proc. of the Genetic and Evolutionary Computation Conference*, pages 59–66, New York, ACM Press, 2005. DOI: 10.1145/1068009.1068018. 129, 141

[15] D. Ashlock and E. Y. Kim. Fingerprinting: Automatic analysis and visualization of prisoner's dilemma strategies. *IEEE Transaction on Evolutionary Computation*, 12:647–659, 2008. DOI: 10.1109/TEVC.2008.920675. 35, 49, 51

[16] D. Ashlock and E. Y. Kim. The impact of varying resources available to iterated prisoner's dilemma agents. In *Proc. of the IEEE Symposium on the Foundations of Computational Inteligence (FOCI)*, 2013. DOI: 10.1109/foci.2013.6602456. 47, 49

[17] D. Ashlock and E. Y. Kim. Techniques for analysis of evolved prisoner's dilemma strategies with fingerprints. In *Proc. of the Congress on Evolutionary Computation*, volume 3, pages 2613–2620, Piscataway, NJ, IEEE Press, 2005. DOI: 10.1109/cec.2005.1555022. 22, 130

[18] D. Ashlock and E. Y. Kim. Prisoner's dilemma: The payoff values matter. In *Proc. of the IEEE Conference on Computational Intelligence in Games*, pages 219–226, Piscataway, NJ, IEEE Press, 2010. DOI: 10.1109/ITW.2010.5593352. 112

[19] D. Ashlock, E. Y. Kim, and W. Ashlock. Fingerprint analysis of the noisy prisoner's dilemma using a finite state representation. *IEEE Transactions on Computational Intelligence and AI in Games*, 1(2):157–167, 2009. DOI: 10.1109/tciaig.2009.2018704. 34, 35, 36, 39, 42, 44, 49, 93, 107

[20] D. Ashlock, E. Y. Kim, and L. Guo. Multi-clustering: Avoiding the natural shape of underlying metrics. In C. H. Dagli et al., Ed., *Smart Engineering System Design: Neural Networks, Evolutionary Programming, and Artificial Life*, volume 15, pages 453–461, ASME Press, 2005. 35

[21] D. Ashlock, E.Y. Kim, and N. Leahy. Understanding representational sensitivity in the iterated prisoner's dilemma with fingerprints. *IEEE Transactions on Systems, Man, and Cybernetics–Part C: Applications and Reviews*, 36(4):464–475, 2006. DOI: 10.1109/tsmcc.2006.875423. 19, 34, 35, 43, 45, 46, 47, 62, 63, 64, 129, 141

[22] D. Ashlock, C. Kuusela, and N. Rogers. Hormonal systems for prisoners dilemma agents. In *Proc. of the IEEE Conference on Computational Inteligence in Games*, pages 63–70, 2011. DOI: 10.1109/cig.2011.6031990. 155, 158

[23] D. Ashlock and C. Lee. Characterization of extremal epidemic networks with diffusion characters. In *Proc. of the IEEE Symposium on Computational Intelligence in Bioinformatics and Computational Biology*, pages 264–271, 2008. DOI: 10.1109/cibcb.2008.4675789. 93

[24] D. Ashlock and C. Lee. Diffusion characters: Breaking the spectral barrier. In *Proc. of the 21st Canadian Conference on Electrical and Computer Engineering*, pages 847–850, 2008. DOI: 10.1109/ccece.2008.4564655. 92

[25] D. Ashlock and J. Mayfield. Acquisition of general adaptive features by evolution. In *Proc. of the 7th Annual Conference on Evolutionary Programming*, pages 75–84, New York, Springer, 1998. DOI: 10.1007/bfb0040761. 70, 74, 75, 78, 93, 127, 134, 141

[26] D. Ashlock and B. Powers. The effect of tag recognition on non-local adaptation. In *Proc. of the Congress on Evolutionary Computation*, volume 2, pages 2045–2051, Piscataway, NJ, IEEE Press, 2004. DOI: 10.1109/cec.2004.1331148. 35, 91, 127

[27] D. Ashlock and C. Richter. The effect of splitting populations on bidding strategies. In *Proc. of the Genetic Programming Conference*, pages 27–34, Cambridge, MA, MIT Press, 1997. 102, 107, 113

[28] D. Ashlock and N. Rogers. A model of emotion in the prisoner's dilemma. In *Proc. of the IEEE Symposium on Computational Intelligence in Bioinformatics and Computational Biology*, pages 272–279, 2008. DOI: 10.1109/cibcb.2008.4675790. 102

[29] D. Ashlock and J. Schonfeld. Nonlinear projection for the display of high dimensional distance data. *Evolutionary Computation. The 2005 IEEE Congress on*, 3:2776–2783, 2006. DOI: 10.1109/cec.2005.1555043. 51

[30] D. Ashlock, M. D. Smucker, E. A. Stanley, and L. Tesfatsion. Preferential partner selection in an evolutionary study of prisoner's dilemma. *Biosystems*, 37:99–125, 1996. DOI: 10.1016/0303-2647(95)01548-5. 30, 35, 49

[31] D. A. Ashlock and E. Y. Kim. Fingerprint analysis of the noisy prisoner's dilemma. In *Proc. of the Congress on Evolutionary Computation*, pages 4073–4080, 2007. DOI: 10.1109/cec.2007.4425002. 107, 129, 132, 134, 142

[32] D. A. Ashlock and J. Schonfeld. Depth annotation of RNA folds for secondary structure motif search. In *Proc. of the IEEE Symposium on Computational Intelligence in Bioinformatics and Computational Biology*, page 3845, 2005. DOI: 10.1109/cibcb.2005.1594896. 130

[33] D. Ashlock and M. Joenks. Isac lists, a different representation for program induction. In *Genetic Programming, Proceedings of the 3rd Annual Genetic Programming Conference*, pages 3–10, San Francisco, CA, Morgan Kaufmann, 1998. 16

[34] W. Ashlock. Why some representations are more cooperative than others for prisoner's dilemma. In *IEEE Symposium on the Foundations of Computational Intelligence*, pages 314–321, Piscataway, NJ, IEEE Press, 2007. DOI: 10.1109/foci.2007.372186. 24, 65

[35] W. Ashlock. Evolving diverse populations of prisoner's dilemma strategies. In *Proc. of the Congress on Evolutionary Computation*, pages 1625–1632, Piscataway, NJ, IEEE Press, 2008. DOI: 10.1109/cec.2008.4631009. 43, 70, 74

[36] W. Ashlock. *Using Priosner's Dilemma Fingerprints to Analyse Evolved Strategies*. Master's thesis, University of Guelph, 2008. 128, 130, 132

[37] W. Ashlock and D. Ashlock. Changes in prisoner's dilemma strategies over evolutionary time with different population sizes. In *Proc. of the Congress On Evolutionary Computation*, pages 1001–1008, Piscataway, NJ, IEEE Press, 2006. DOI: 10.1109/cec.2006.1688322. 11, 19, 44, 70, 73, 136, 138, 141

[38] R. Axelrod. *The Evolution of Cooperation*. Basic Books, New York, 1984. DOI: 10.1126/science.7466396. 3, 129

[39] R. Axelrod and W. D. Hamilton. The evolution of cooperation. *Science*, 211:1390–1396, 1981. DOI: 10.1126/science.7466396. 68, 90, 114, 129

[40] W. Banzhaf, P. Nordin, R. E. Keller, and F. D. Francone. *Genetic Programming: An Introduction on the Automatic Evolution of Computer Programs and its Applications*. Morgan Kaufmann, San Francisco, CA, 1998. 14

[41] L. Barlow and D. Ashlock. Varying decision inputs in prisoner's dilemma. In *Proc. of the IEEE Conference on Computational Intelligence in Bioinformatics and Computational Biology*, pages 1–8, 2015. DOI: 10.1109/cibcb.2015.7300295. 102

[42] J. A. Brown. *Regression and Classification from Extinction*. Ph.D. thesis, University of Guelph, 2014. 113

[43] D. Challet and Y. Zhang. Emergence of cooperation and organization in an evolutionary game. *Physica A*, 246:407–418, 1997. DOI: 10.1016/s0378-4371(97)00419-6. 116

[44] D. Challet and Y. Zhang. On the minority game: Analytical and numerical studies. *Physica A*, 256:514–532, 1998. DOI: 10.1016/s0378-4371(98)00260-x. 115, 119, 123

[45] X. S. Chen, Y. S. Ong, and M. H. Lim. Research frontier: Memetic computation—past, present and future. *IEEE Computational Intelligence Magazine*, 2(5):24–36, 2010. DOI: 10.1109/MCI.2010.936309. 153

[46] S. Y. Chong and X. Yai. Self-adapting payoff matrices in repeated interactions. In *Proc. of the IEEE Symposium on Computational Intelligence in Games*, pages 103–110, Piscataway, NJ, IEEE Press, 2006. DOI: 10.1109/cig.2006.311688. 33, 102

[47] S. Y. Chong and X. Yao. Behavioral diversity, choices and noise in the iterated prisoner's dilemma. *IEEE Transaction on Evolutionary Computation*, 9:540–551, 2005. DOI: 10.1109/tevc.2005.856200. 24

[48] R. Croson. Theories of commitment, altruism and reciprocity: Evidence from linear public goods games. *Economic Inquiry*, 45(2):199–216, 2007. DOI: 10.1111/j.1465-7295.2006.00006.x. 116

[49] E. Shiller, D. Ashlock, and C. Lee. Comparison of evolved epidemic networks with diffusion characters. In *Proc. of the IEEE Congress on Evolutionary Computation*, pages 781–788, Piscataway, NJ, IEEE Press, 2011. DOI: 10.1109/cec.2011.5949698. 93

[50] M. Marsili, D. Challet, and R. Zecchina. Statistical mechanics of systems with heterogeneous agents: Minority games. *Physical Review Letters*, 84:1824–1827, 2000. DOI: 10.1103/physrevlett.84.1824. 116

[51] R. Dawes. Social dilemmas. *Annual Review of Psychology*, 31:169–193, 1980. DOI: 10.1146/annurev.ps.31.020180.001125. 114, 118

[52] R. Dawkins. *The Blind Watchmaker*. W. W. Norton and Company Ltd., Cambridge, 1986. 74

[53] M. de Berg, M. van Kreveld, M. Overmans, and O. Schwarzkopf. *Computational Geometry: Algorithms and Applications*. Springer-Verlag, Berlin, 2000. DOI: 10.1007/978-3-540-77974-2. 130

[54] M. Doebeli and C. Hauert. Models of cooperation based on the prisoner's dilemma and the snowdrift game. *Ecology Letters*, 8:748–766, 2005. DOI: 10.1111/j.1461-0248.2005.00773.x. 114

[55] M. Dorigo and L. M. Gambardella. Ant colony system: A cooperative learning approach to the traveling salesman problem. *IEEE Transactions on Evolutionary Computation*, 1(1):53–66, 1997. DOI: 10.1109/4235.585892. 152

[56] D. Doty. Non-loca evolutionary adaptation in grid plants. In *Proc. of the Congress on Evolutionary Computation*, volume 2, pages 1602–1609, Piscataway, NJ, IEEE Press, 2004. DOI: 10.1109/cec.2004.1331087. 93, 127

[57] R. Kummerli et al. Human cooperation in social dilemmas: Comparing the snowdrift game with the prisoner's dilemma. *Proc. of the Royal Society B*, 274:2965–2970, 2007. DOI: 10.1098/rspb.2007.0793. 115

[58] T. Qiu et al. Cooperation in the snowdrift game on directed small-world networks under selfquestioning and noisy conditions. *Computer Physics Communications*, 181(12):2057–2062, 2010. DOI: 10.1016/j.cpc.2010.08.018. 115

[59] K. Foster, T. Wenseleers and F. L. W. Ratnieks. Kin selection is the key to altruism. *Trends in Ecology and Evolution*, 21(2):57–60, 2000. DOI: 10.1016/j.tree.2005.11.020. 102

[60] O. Frank and F. Harary. Cluster inference by using transitivity indices in empirical graphs. *Journal of the American Statistical Association*, pages 835–840, 1982. DOI: 10.2307/2287315. 91

[61] N. Franken and A. P. Engelbrecht. Particle swarm optimization approaches to coevolve strategies for the iterated prisoner's dilemma. *IEEE Transaction on Evolutionary Computation*, 9:562–579, 2005. DOI: 10.1109/tevc.2005.856202. 24

[62] G. W. Greenwood and S. Chopra. A game-theoretic and dynamical-systems analysis of selection methods in coevolution. *IEEE Transactions on Evolutionary Computation*, 9(6):580–602, 2005. DOI: 10.1109/tevc.2005.856203. 87, 114

[63] P. J. F. Groenen and I. Borg. *Modern Multidimensional Scaling: Theory and Applications*. Springer-Verlag, Secaucus, NJ, 2005. DOI: 10.1007/0-387-28981-X. 51

[64] F. Gruau. *Neural Network Synthesis using Cellular Encoding and the Genetic Algorithm*. Ph.D. thesis, Ecole Normale Supérieure de Lyon, France, 1994. 12

[65] C. Hauert and M. Doebeli. Spatial structure often inhibits the evolution of cooperation in the snowdrift game. *Nature*, 428:643–646, 2004. DOI: 10.1038/nature02360. 115

[66] S. Haykin. *Neural Nets, a Comprehensive Foundation*. Macmillan College Publishing, New York, 1994. 17, 151

[67] M. Hemesath. Cooperate or defect? Russian and American students in a prisoner's dilemma. *Comparative Economics Studies*, 176:83–93, 1994. DOI: 10.1057/ces.1994.6. 33

[68] P. F. Hingston, L. C. Barone, and Z. Michalewicz. *Design by Evolution*. Springer, 2010. DOI: 10.1007/978-3-540-74111-4. 153

[69] J. M. Houston, J. Kinnie, B. Lupo, C. Terry, and S. S. Ho. Competitiveness and conflict behavior in simulation of a social dilemma. *Psychological Reports*, 86:1219–1225, 2000. DOI: 10.2466/pr0.86.3.1219-1225. 33

[70] H. Ishibuchi and N. Namikawa. Evolution of iterated prisoner's dilemma game strategies in structured demes under random pairing in game playing. *IEEE Transaction on Evolutionary Computation*, 9:540–551, 2005. DOI: 10.1109/tevc.2005.856198. 24

[71] Y. Jin and B. Sendhoff. Constructing dynamic optimization test problems using the multi-objective optimization concept. In G. R. et al. Raidl, Ed., *Applications of Evolutionary Computing*, volume 3005 of *Lecture Notes in Computer Science*, pages 525–536, Springer Berlin Heidelberg, 2004. DOI: 10.1007/3-540-46004-7. 112

[72] E. D. De Jong. Intransitivity in coevolution. In *Parallel Problem Solving from Nature-PPSN VIII*, pages 843–851, Berlin, Heidelberg, Springer, 2004. DOI: 10.1007/978-3-540-30217-9_85. 87

[73] C. H. Roth Jr. *Fundamentals of Logic Design*. Thomson-Engineering, 2004. 45

[74] K. Lindgren and M. G. Nordahl. Evolutionary dynamics of spatial games. In *Proc. of the Oji International Seminar on Complex Systems: From Complex Dynamical Systems to Sciences of Artificial Reality*, pages 292–309, Elsevier North-Holland, Inc., New York, 1994. DOI: 10.1016/0167-2789(94)90289-5. 34, 102, 151

[75] J. Kennedy and R. Eberhart. Particle swarm optimization. In *Proc. of IEEE International Conference on Neural Networks*, volume IV, pages 1942–1948, 1995. DOI: 10.1109/icnn.1995.488968. 152

[76] E. Y. Kim. *Analysis of Game Playing Agents with Fingerprints*. Ph.D. thesis, Iowa State University, 2005. 30, 141

[77] Q. Chen and K. Flick. Var genes, PfEMP1 and the human host. *Molecular and Biochemical Parasitology*, 134:3–9, 2004. DOI: 10.1016/j.molbiopara.2003.09.010. 84

[78] P. Legendre and L. Legendre. *Numerical Ecology*, 2nd ed., Elsevier Science, Amsterdam, 1998. 51

[79] J. Li, P. Hingston, and G. Kendall. Engineering design of strategies for winning iterated prisoner's dilemma competitions. *IEEE Transactions of Computational Intelligence and AI in Games*, pages 348–360, 2011. DOI: 10.1109/tciaig.2011.2166268. 68

[80] K. Lindgren. Evolutionary phenomena in simple dynamics. In *Artificial Life II*, pages 295–312, 1992. 151

[81] M. Loreau and C. de Mazancourt. Biodiversity and ecosystem stability: A synthesis of underlying mechanisms. *Ecology Letters*, 16:106–115, 2013. DOI: 10.1111/ele.12073. 74

[82] C. Hauert, M. Doebeli, and T. Killingback. The evolutionary origin of cooperators and defectors. *Science*, 306:859–861, 2004. DOI: 10.1126/science.1101456. 114, 126

[83] A. Chakraborti, M. Sysi-Aho, and K. Kaski. Intelligent minority game with genetic crossover strategies. *European Physics Journal B*, 34:373–377, 2003. DOI: 10.1140/epjb/e2003-00234-0. 117

[84] J. F. Miller and S. L. Smith. Redundancy and computational efficiency in cartesian genetic programming. *IEEE Transaction on Evolutionary Computation*, 10(2):167–174, 2006. DOI: 10.1109/tevc.2006.871253. 19

[85] J. H. Miller. The coevolution of automata in the repeated prisoner's dilemma. *Journal of Economic Behavior and Organization*, 29(1):87–112, 1996. DOI: 10.1016/0167-2681(95)00052-6. 24, 72, 132

[86] J. F. Miller and P. Thomson. Cartesian genetic programming. In *Proc. of the European Conference on Genetic Programming*, pages 121–132, Springer-Verlag, London, UK, 2000. DOI: 10.1007/978-3-642-17310-3_2. 15

[87] J. Nash. Two-person cooperative games. *Econometrica*, 21(1):128–140, 1953. DOI: 10.2307/1906951. 101

[88] M. J. Osborne and A. Rubinstein. *A Course in Game Theory*. The MIT Press, 1994. 88

[89] J. Pacheco and F. Santos. Network dependence of the dilemmas of cooperation. In *AIP Conference Proceedings*, volume 776, pages 90–100, AIP Publishing LLC, 2005. DOI: 10.1063/1.1985380. 115

[90] D. Parrott and X. Li. Locating and tracking multiple dynamic optima by a particle swarm model using speciation. *Evolutionary Computation, IEEE Transactions on*, 10(4):440–458, 2006. DOI: 10.1109/tevc.2005.859468. 112

[91] R. Riolo, R. Manuca, Y. li, and R. Savit. The structure of adaptive competition in minority games. *Physica A*, 282:559–608, 2000. DOI: 10.1016/s0378-4371(00)00100-x. 122

[92] R. Manuca, R. Savit, and R. Riolo. Adaptive competition, market efficiency and phase transitions. *Physical Review Letters*, 82(10):2203–2206, 1999. DOI: 10.1103/physrevlett.82.2203. 119

[93] S. I. Resnick. *Adventures in Stochastic Processes*. Birkhauser, Boston, 1992. DOI: 10.1007/978-1-4612-0387-2. 19, 21

[94] R. G. Reynolds. An introduction to cultural algorithms. In *Proc. of the 3rd Annual Conference on Evolutionary Programming*, pages 131–139, World Scientific Publishing, 1994. 152

[95] G. Rezaei and M. Kirley. The effects of time-varying rewards on the evolution of cooperation. *Evolutionary Intelligence*, 2(4):207–281, 2009. DOI: 10.1007/s12065-009-0032-1. 33, 102

[96] D. Roy. Learning and the theory of games. *Journal of Theoretical Biology*, 204:409–414, 2000. DOI: 10.1006/jtbi.2000.2022. 33

[97] T. P. Runarsson, S. Samothrakis, S. Lucas, and D. Robles. Coevolving game-playing agents: Measuring performance and intransitivities. *IEEE Transactions on Evolutionary Computation*, 17:213–226, 2012. DOI: 10.1109/tevc.2012.2208755. 91, 99

[98] Y. G. Seo, S. B. Cho, and X. Yao. The impact of payoff function and local interaction on the n-player iterated prisoner's dilemma. *Knowledge and Information Systems*, 2:461–478, 2000. DOI: 10.1007/pl00011652. 33, 102

[99] K. Sigmund and M. A. Nowak. Evolutionary game theory. *Current Biology*, 9(14):R503–505, 1999. DOI: 10.1016/s0960-9822(99)80321-2. 33

[100] A. Stanley, D. Ashlock, and L. Tesfatison. Iterated prisoner's dilemma with choice and refusal. *Economic Report*, 30, 1992. 25

[101] E. A. Stanley, D. Ashlock, and L. Tesfatsion. Iterated prisoner's dilemma with choice and refusal. In Christopher Langton, Ed., *Artificial Life III*, volume 17 of *Santa Fe Institute Studies in the Sciences of Complexity*, pages 131–176, Addison-Wesley, Reading, 1994. 152

[102] J. H. van Lint and R. M. Wilson. *A Course in Combinatorics*. Cambridge University Press, Cambridge, UK, 2001. DOI: 10.1017/cbo9780511987045. 23

[103] M. Hu, W. du, X. Cao, and W. Wang. Asymmetric cost in snowdrift game on scale-free networks. *European Physics Letters*, 87:60004p1–6004p6, 2009. DOI: 10.1209/0295-5075/87/60004. 115

[104] Y. Wu, W. Yang, B. Wang, and Y. Xie. Searching good strategies in evolutionary minority game using variable length genetic algorithm. *Physica A*, pages 583–590, 2004. DOI: 10.1016/j.physa.2004.03.065. 117

[105] K. A. Yurkonis, B. J. Wilsey, and K. A. Moloney. Initial species pattern affects invasion resistance in experimental grassland plots. *Journal of Vegetation Science*, 23(1):4–12, 2012. DOI: 10.1111/j.1654-1103.2011.01331.x. 74

[106] D. Zheng, C. Chan, H. Yin, and P. Hui. Cooperative behavior in a model of evolutionary snowdrift games witn n-person interactions. *EPL*, 80:18002–18005, 2007. DOI: 10.1209/0295-5075/80/18002. 115, 124, 126

[107] G. Greenwood and D. Ashlock. Evolutionary games and the study of cooperation: Why has so little progress been made? *Proc. of the IEEE Congress on Evolutionary Computation*, pages 680–687, 2012. 87

Authors' Biographies

EUN-YOUN KIM

Dr. Eun-Youn Kim is a mathematician with a background in graph theory and evolutionary game theory, having worked extensively on the issue of representation in the design of game-playing agents. Dr. Kim received her Ph.D. in mathematics from Iowa State University. She has been employed at the National Institute of Mathematical Sciences and the Korea Institute of Bioscience and Biotechnology in Korea.

Dr. Kim's research interests include broad interests in evolutionary computation applied to game playing and network modeling in biological systems. She is a member of the IEEE Computational Intelligence Societies technical committee on games and has helped organize international games conferences.

Dr. Kim is currently employed by the department of basic science in Hanbat National University in South Korea.

DANIEL ASHLOCK

Dr. Daniel Ashlock is a professor of mathematics at the University of Guelph in Ontario, Canada. Dr. Ashlock received his Ph.D. in mathematics from Caltech with a focus in algebraic combinatorics. He was employed at Iowa State University before moving to Canada.

Dr. Ashlock works on representation issues in evolutionary computation including games, optimization, bioinformatics, and theoretical biology. He holds the Bioinformatics Chair in the Department of Mathematics and Statistics at Guelph and serves on the editorial board of the IEEE Transactions on Evolutionary Computation, the IEEE Transactions on Games, The IEEE/ACM Transactions on Bioinformatics and Computational Biology, Biosystems, and Game and Puzzle Design.

Dr. Ashlock serves on the IEEE Computational Intelligence Societies technical committees on games and bioinformatics and biomedical engineering.

Printed in the United States
by Baker & Taylor Publisher Services